Pipejacking and Microtunnelling

PIPEJACKING AND MICROTUNNELLING

By

James C. Thomson, C. Eng., Eur. Ing.
Chairman
Jason Consultancy Group
London - Geneva - Washington DC

CRC Press
Taylor & Francis Group
Boca Raton London New York

CRC Press is an imprint of the
Taylor & Francis Group, an informa business
A TAYLOR & FRANCIS BOOK

CRC Press
Taylor & Francis Group
6000 Broken Sound Parkway NW, Suite 300
Boca Raton, FL 33487-2742

First issued in paperback 2019

ISBN-13: 978-0-7514-0102-8 (hbk)
ISBN-13: 978-0-367-86472-9 (pbk)

A Catalogue record for this book is available from the British Library

Library of Congress Cataloging-in-Publication Data available

Publisher's Note
The publisher has gone to great lengths to ensure the quality of this reprint but points out that some imperfections in the original may be apparent

Visit the Taylor & Francis Web site at
http://www.taylorandfrancis.com

and the CRC Press Web site at
http://www.crcpress.com

Foreword

The Author draws upon his wide experience over many years to recount the several features which may combine to favour pipejacking techniques for the installation of underground services. The simple traditional operation of thrust-boring is seen to have blossomed into a highly diversified genera of specialised techniques to satisfy differing requirements of use, alignment, surface access and, the quintessential factor, the ground.

The main purpose of this book is to set out the state of the art, in particular to describe the merits and applications of specialist plant and equipment currently used. Since one of the objectives of pipejacking is to minimise the level of activity at the leading end of the tunnel, overall reliability and the appropriate matching of plant for excavation, spoil removal and control of line must be essential criteria for success.

This book may also be interpreted as partly biographical of the Author's career. This has clearly included searing experiences of poor contract administration which cause him to turn his fire on *traditional* Conditions of Contract. There are features here which call for broader assessment. Forms of contract which include the independent Engineer depend for success upon administration by an Engineer who is professionally motivated and well skilled in the construction techniques of the Contract. It is here that many of the contractual problems in pipejacking have arisen since too often those who have performed the role of Engineer may have been highly experienced in the use of the particular service but not with the engineering of its installation. It is moreover a fact that in tunnelling many of the economic benefits of innovation have been derived through the *traditional* route, the incentives being those of professional motivation, including concern for the interests of the Client.

All civil engineering needs to take account of the continuity through the several processes which link the initial demand to its ultimate satisfaction. The contractual relationships should respect the particular contributions to be expected from those concerned. Within this framework, the sharing of risk is to be encouraged to minimise cost, to maximise clarity in the acceptance of responsibility and to avoid the *brittle* contract, so often the cause of loss for all parties and for the avoidable subsequent litigation. It deserves mention that risk-sharing has long been practised in the United Kingdom, and through traditional Contract forms (see, for example, *Tunnelling — Improved Contract Practices CIRIA 1978*). The down-side for Target Contracts in tunnelling includes the greater cost of administration of accountancy and the issues attending the variation in the Target Value — where embattled adversarial positions are taken up!

This book merits study by those interested to increase their understanding of the techniques and benefits of pipejacking. The decision whether or not to prefer pipejacking must also entail the comparison of the merits of other forms of tunnelling for a particular application. For the more difficult circumstances of ground and alignment there is need to bear in mind, for example, the potential difficulties in the recovery of a pipe-jacked shield which is damaged or off-line. Most particularly, site investigation needs to be co-ordinated with the decision process, since frequently the preference for the method of construction cannot be confirmed until the site investigation is in hand. Meanwhile the site investigation must address the requirements of all possible contenders.

Pipejacking will unquestionably serve for an increasing share of new underground services and, in somewhat different forms, for the renewal of existing services. It is well, therefore, that all who may become involved in the techniques should be familiar with the options, the opportunities and the pitfalls to be avoided. This book provides a good starting point.

Sir Alan Muir Wood FRS F.Eng. FICE

Preface

In 1892, the Minutes of the 10th Annual Convention of The New England Roadmasters Association included perhaps the earliest reference to pipejacking:

> *Mr W.E. Clark described a method which he had followed in putting a wrought iron pipe through an earth bank by means of hydraulic jacks. The first section of pipe was laid in a trench cut into the side of the bank, and the remaining sections were put on the ends, one after another, and the whole was pushed forwards by the power of the jacks, somewhat after the manner in which tunnelling in soft ground has, of late years, been so successfully practiced.*

The fact that jacking methods have been employed for over 100 years may come as a surprise even to engineers in the trenchless technology business. To those who still remain sceptical about these *new-fangled unproven methods*, this documented reference should prompt more considered attention to what is now a technically sophisticated, multimillion dollar industry. Pipejacking and the smaller scale microtunnelling, as part of this industry, are established as effective non-disruptive ways of installing pipes and cables underground. In an environment-conscious world, where the focus of pipe laying and rehabilitation is in urban areas, the burgeoning demand for tunnelled installations can only continue to grow.

Perhaps one reason why many engineers and contractors still have little understanding of jacking methods, and in some cases a continuing resistance to them, is that there is a remarkable lack of any comprehensive literature on the subject. Two valuable books produced so far have been written in German and naturally focus on German practice and usage, even though Dr Stein's book has been translated into English. A great many papers and articles have been published, including my own, but these have referred to specific problems or particular areas of work. The only other references appear as short sections devoted to pipejacking methods in several books on general tunnelling; at best these have tended to be patronising and at worst inaccurate.

My objectives in writing this book are therefore threefold. The first is to provide a review of pipejacking and microtunnelling technology and its application, reflecting international experience as far as possible. The second is to offer some guidance on the use of these techniques and to persuade engineers that these can offer economic as well as technical benefits when used correctly in appropriate circumstances. It is not intended to be a definitive work on jacking, as to give adequate cover to all its aspects and related technologies would require several volumes and would mean crossing into many disciplines, most of which are well documented elsewhere. My third objective is to celebrate 100 years of pipejacking and to pay tribute to its forgotten pioneers.

To these, I should add a fourth of a more personal nature. Having worked with pipejacking since 1958, the greater part of my professional life, I felt it was time to gather together my hard won experience and knowledge of its present technical and commercial status. As I have been preaching the gospel of trenchless technology since long before it became *fashionable*, I now find it extremely satisfying to see so many converts to the cause.

In the 1960s and 1970s, my ideas on soft ground tunnelling were treated with some scorn by the tunnelling establishment — both consultants and contractors. On one occasion the doyen of the industry, Sir Harold Harding, explained to me that I, like his wife, failed to understand the fundamental principles of tunnelling. Apparently, whilst on the beach with his family, he had been playing with a tin can in the sand and water. As she watched, his wife had observed, "Harold, why don't you push in your tunnel linings with a lot of cans rather than building the lining behind the shield?" In simple terms he had given her all the engineering reasons why this was not possible. This straight-faced account was followed by a wry smile and the admission, "I hope you appreciate that the work you have been doing requires me to go back and tell her she might have been right."

My own wife, Ann, a consultant in Business Communications and technical editor, has been a partner in writing this book. It is she who has taken on the daunting task of turning an engineer's ideas into some semblance of order and readability. If justice is to be done, this book should be attributed to our joint authorship.

Contributions from my colleagues in Jason Consultants and my many friends in the industry are also integrated into these chapters. I would like to thank all these people, too numerous to mention by name, for their generosity and support. Photographs, drawings, tables and examples kindly offered by many individuals, companies and organisations I have tried to acknowledge suitably in the text. However, I am particularly indebted to Dr. Chris Rogers of Loughborough University for his erudite contribution on theory and analysis related to permanent and temporary loads on pipes and to face stability in Chapters 8, 9, and 10.

If you look up "Boring" in the Yellow Pages of the London Telephone Directory, it says, "*see* Civil Engineers". My family and friends might be inclined to accept this classification of a civil engineer who has spent most of his working life driving holes through the ground. However, as anybody who has had the privilege of working below ground well knows, life is challenging and exciting, not boring or routine. My earnest hope is that this book will not be a bore to the reader, but an opportunity to share my enthusiasm for jacking and to appreciate the versatility of application it offers.

J.C.T.

Contents

1 Development of trenchless technology

1.1 Recent growth of trenchless technology

Trenchless technology broadly brings under one banner a variety of non-disruptive techniques for installing, replacing or renovating underground pipes and cables without open-cut excavation. The term was introduced as recently as the early 1980s. Through the efforts of the International Society for Trenchless Technology (ISTT) and information disseminated at No-Dig Conferences, *trenchless technology* has become accepted as a specific construction term. Pipejacking, as both a principle and a technique, is a trenchless method of pipelaying that existed long before this term.

The pipelaying industry is now having to move away from its traditional ways of working. Not only is it having to adapt to changes in client needs, but also to changing attitudes of both publics and governments (local and central) as they respond to the wider social, political and economic changes around them.

In industrialised countries, pipelaying is no longer primarily concerned with installing new networks, as most households are already connected to sewers, gas, water, electricity and telecommunication services. Work is now more usually focused in established urban areas where ageing utility networks, often well past their design life, are in need of replacement or rehabilitation. By definition, these are the areas with the densest population and the greatest congestion. The whole question of how pipes are installed or renovated is understandably being thrown into sharper relief.

Public and government resistance is hardening against the disruption of traffic, commerce and community well-being caused by open-cut methods of working. People are now realising that traditional open trenching inflicts real costs on communities, as well as inconvenience. The disruption factor of pipelaying is under scrutiny and from now on it will be an element of installation programmes which must be accounted for. To meet the challenge of changing needs and stricter requirements, new and often more sophisticated trenchless techniques of pipelaying have been introduced, and existing ones refined. As these techniques are more widely accepted as viable alternatives to open-cut working, trenchless systems are competing with traditional installation methods — not just on technical grounds, but also for good economic reasons. With greater production and increased utilisation rates, prices for trenchless solutions have become directly competitive.

1.1.1 Categories of trenchless technology

The many types of trenchless technology now available are generally categorised into three main groups according to their application, as shown in Table 1.1.

Table 1.1 Categories of trenchless technology

Category	Function
New installation	Putting a new pipe on line
On-line replacement	Putting a new pipe on an old line
Renovation	Lining an old pipe

The following charts show the types of trenchless technology available and their applications. Table 1.2 summarises installation methods for all types of pipe, for gravity sewers and pressure pipe. Table 1.3 lists the methods of installing sewers and pressure pipes for cased crossings. Trenchless methods for on-line replacement and for structural renovation are summarised in Tables 1.4 and 1.5 respectively, gravity sewer and pressure pipe applications again being shown separately.

Table 1.2 Trenchless methods used for new installation

Gravity sewers	Pressure pipes
Pipe jacking > 900 mm (36 in)	Directional drilling
Microtunnelling < 900 mm (36 in)	Guided boring
	Impact moling
	Rod pushing

Table 1.3 Trenchless methods for cased crossing

Gravity sewers	Pressure pipes
Pipe jacking	Directional drilling
Microtunnelling	Guided boring
Auger boring	**Pipe jacking**
Impact ramming	**Microtunnelling**
	Auger boring
	Impact ramming

Table 1.4 Trenchless methods for on-line replacement

Gravity sewers	Pressure pipes
Bursting	Bursting
Excavation	Splitting
	Extraction

Table 1.5 Trenchless methods for structural renovation

Linings	Coatings
Sliplining	Reinforced shotcrete
Pipes (fusion-welded)	Cast-in-place concrete
Sliplining (short pipes)	Spray-on linings
Cured-in-place	
Deformed pipe	
Spiral-wound pipe	
Segmental linings	

1.2 Defining pipejacking and microtunnelling

1.2.1 Pipejacking

Pipejacking describes a specific trenchless technique for installing pipes. *Pipejacking* is also used as a more general term for a principle which is fundamental to a number of trenchless techniques. It is important to differentiate between the principle and the technique.

The principle of using hydraulic rams to push up pipe sections to line the hole formed by a cutting head, or shield, applies to auger boring and microtunnelling as well as to the pipejacking method itself. The principle of jacking pipes is the same whether one is installing a 100 mm auger bore casing or driving a concrete underpass for vehicles. Figures 1.1 to 1.3 show how the pipejacking principle is applied to auger boring, microtunnelling, and pipejacking. Figure 1.4 is an example of a jacked underpass. The size range varies from 100 mm for the auger bore to 20 m by 7 m for the underpass.

Pipejacking is the name given to the technique of installing pipe of man-entry size by pushing sections from the drive shaft into the hole formed by the cutting shield to provide a continuous tunnel lining. These sections do not have to be circular and can be of almost any dimensions. The cutting shield external dimension can be sized to conform closely to the external diameter of the pipe being pushed; this eliminates or reduces to a minimum the annular space created. Diameters considered man-entry vary slightly from one country to another, but 900 mm has become a widely accepted lower limit.

Pipejacking was initially developed for installing casings for crossings and is still widely used for this purpose. As techniques and skills improved, it evolved into a method capable of installing pipes direct without a casing, particularly gravity sewers. By so doing, pipejacking has moved into the traditional small diameter tunnelling market. Here it competes with alternative methods of installation, in particular with segment linings, which are erected immediately behind the shield as it advances. The shields and tunnel boring machines (TBMs) used at the leading end of the tunnel are virtually the same for jacking and segment tunnelling. The only difference is in the way the tunnel is lined.

Figure 1.1 Auger boring rig

Figure 1.2 Microtunnelling

Figure 1.3 Pipejacking

Figure 1.4 Jacked underpass (by courtesy of Cementation Projects Ltd)

1.2.2 Microtunnelling

Microtunnelling is the term for pipejacking in smaller diameters, as its name implies. Microtunnelling equipment is used to install pipes with an internal diameter less than 900 mm, i.e. diameters generally accepted as too small for safe man-entry working. However, the upper limit of 900 mm is not sacrosanct. The Japanese consider an internal diameter of 800 mm large enough for man-entry working, whereas in Europe there is a move towards raising the microtunnelling limit to 1000 mm, particularly for longer drives. Whatever the precise upper limit, microtunnelling is a method of installing pipes to a predetermined line and level by remote control. Excavation, spoil removal and shield steering are all remotely controlled.

The difference between pipejacking and microtunnelling is therefore chiefly a matter of size and not a question of whether the system is or is not remotely controlled. As far as categorising equipment is concerned, the division is meaningless. A manufacturer may have a range of remotely controlled machines of the same design ranging from 500 mm to 1500 mm or greater. Remote-control methods are increasingly being selected for installing pipes of 2000 mm and less.

The predominant application for microtunnelling is the direct installation of gravity sewer pipes, but other forms of pipe and duct are also direct-installed in this way, although still a relatively small proportion. In some circumstances microtunnelling can be an effective means of installing duct at crossings.

A spin-off from the development of remotely controlled machines for new installation has been their use in the on-line replacement of sewers. Microtunnelling equipment has been adapted to break up and excavate an existing sewer and to lay a replacement pipe all in one operation.

1.3 A historical perspective

1.3.1 Pipejacking

The pipejacking principle was thought to have first been used in Roman times. Excavations in the Middle East suggest that a wooden pipe was jacked underground with a series of levers and tapped into the side of a Roman aqueduct to gain illicit access to water. The motive was different, but the objective was the same: to lay a pipe without disturbing the surface.

Early pipejacking in the USA. The first recorded description of the pipejacking method seems to have been in 1892 [1]. The first records of pipejacking jobs were for installations done in the USA between 1896 and 1900 for the Northern Pacific Railroad Company [2] [3]. Gradually it became standard practice for a number of railroad companies to jack cast iron pipes under rail tracks. Further records show pipes were jacked under the Great Western Railway at Ingalton, Illinois, in 1911 and for Southern Pacific in California in 1915.

Augustus Griffin is credited with the early popularisation of the jacking method (see Figure 1.5). It was while working on irrigation schemes in California, between 1906 and 1918, that Griffin developed techniques for jacking cast iron pipe culverts under rail tracks. Figures 1.6 and 1.7 are taken from one of Griffin's articles [4].

Corrugated steel pipes started to replace cast iron in the early 1920s. Pipejacking accounted for some 830 jobs between 1922 and 1947, totalling 16,800 m (56,000 ft) with diameters from 700 mm to 2400 mm (28-96 in). In his paper published in 1948, Jacob Feld [5] lists in a table some of these installations, which in some cases achieved drives of 60 m (200 ft) or more. Figure 1.8 shows an early crossing under an embankment.

By the late 1920s, concrete pipes were being used. A paper published in 1942 for the American Concrete Pipe Association [6] refers to some jobs using concrete pipes in the 20-year period from 1927. Pipe sizes ranged from 750 mm to 2400 mm (30-96 in). In the 1930s, Northern Pacific standardised the use of concrete jacking pipes with internal diameters from 108 mm to 1800 mm (42-72 in). Of particular note was the jacking in 1945 of a concrete pipe, 2400 mm (96 in), for a pedestrian subway under the Richmond, Fredricksburg and Potomac railroad at the Potomac yards, Alexandria. Cover from the base of the railway was a scant 750 mm (2 ft 6 in).

Records reveal that working methods were similar to a simple hand-mined jacking job of today. Figure 1.9 shows the plan of a jacking set-up while Figures 1.10 and 1.11 are photographs of typical operations. It would appear that although many, but by no means all, the pipes were fitted with a leading steel cutting edge, no separate adjustable shield was employed. Steering in good ground was done by excavating in front of the pipe to the course required. The devices used for jacking were probably whatever was available from other applications; reference can be found to screw jacks, ratchet jacks, air piston cylinders and hydraulic jacks. Hydraulic jacks had the advantage of allowing jacking pressures to be recorded. Concrete pipe in 1200 mm (4 ft) lengths with tongue-and-groove joints was common, but 2400 mm (8 ft) lengths became increasingly popular from the late 1930s.

Excavation was mainly by shovel and pick, helped by pneumatic clay spades in tougher soils with air hammers to break out boulders. A suggested alternative was to heat the rock with oil torches and to apply water. Progress varied according to conditions, but in 1937 a 60 m (200 ft) drive of 900 mm (36 in) corrugated pipe took 20 days to install, using three shifts. Daily progress recorded on jacking concrete pipes in the late 1930s varied from 1.5 to 2.1 m (5 ft to 7 ft) a day for 1500 mm and 1800 mm (60 in and 72 in) pipe and from 4.3 m to 5.2 m (14 to 17 ft) a day for 1200 mm (48 in) driving in dewatered sandy soil.

In 1936 the installation of a 1500 mm (60 in) concrete pipe 30 m (96 ft) long in a slag dump was done for $84 per metre ($25.55 per ft). A 1350 mm (54 in) concrete pipe 24 m (79 ft) long in good soil cost $60 per metre ($18.45 per ft) in 1934. The labour cost for jacking work was typically put between $13-$20 per metre ($4 to $6 per ft). These rates seem to have been current from the late 1920s to the early 1940s.

Figure 1.5 Augustus Griffin (Born 1883 in Visalia, California, died 1946, Calgary.

Figure 1.6 Jacking bell-and-spigot concrete pipe. (photo by courtesy of *Engineering News Record*, 1924)

Figure 1.7 Jacking 36 in. corrugated pipe (photograph by courtesy of Engineering News Record, 1924)

Figure 1.8 Jacking corrugated pipe through an embankment (photo by courtesy of Contech Inc.)

Even though pipejacking was mainly for installing pipelines and ducts at crossings under railways and highways, papers of the time discussed the advantages of installing pipe for drains and other pipelines directly in the ground, without a casing. The benefit of minimum disturbance to traffic was emphasised even then. Jacob Feld [5] in his paper states that, if backfilling and restoration costs were taken into account, it was cheaper to jack pipe under the railroad than to install it by open-cut methods, even without allowing for the cost of traffic delays.

Similar methods continued within the USA with relatively little development of technique until well into the 1960s [7]. Manufacturers' literature in 1964 still offered fixed shields, some with hoods and shelves. However, some progress was made. Multi-point hydraulic long-stroke jacking systems were produced with capacities up to 400 tons. Much longer lengths, up to 152 m (500 ft) were being driven.

Until quite recently, the USA seems neither to have fully appreciated nor to have capitalised on the potential of its own invention. Much pipejacking work is still for crossings and often undertaken by contractors with unsophisticated equipment and unskilled operatives. One reason for this may be that past problems with pipejacking have given some authorities and engineers a jaundiced view of this method.

Pipejacking develops in Europe and Japan. Prior to World War II, the pipejacking technique was used on an isolated basis in various countries, among them the UK, Germany and Japan. Jacking concrete pipes is described in a paper of 1934 [8] and records exist of concrete pipe being jacked at Nurenburg in 1935 [9].

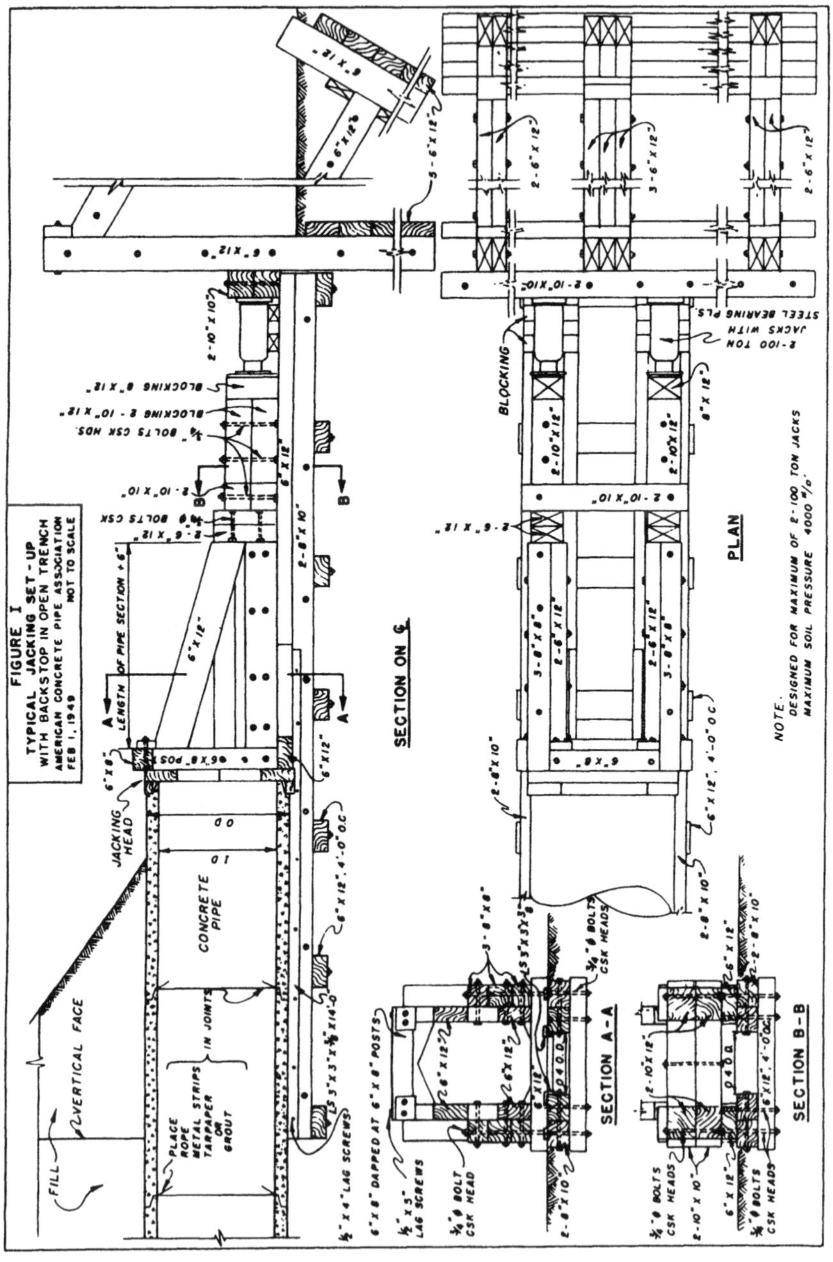

Figure 1.9 Drive pit arrangements (drawing, *Jacking Reinforced Concrete Pipe Lines,* Blue Book, page 20, by courtesy of the American Concrete Pipe Association)

Figure 1.10 Jacking pit, Memphis (photograph from *Jacking Reinforced Concrete Pipe Lines*, Blue Book, by courtesy of American Concrete Pipe Association)

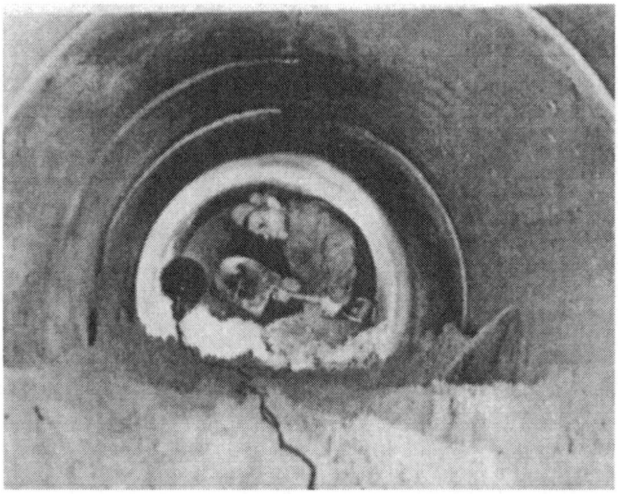

Figure 1.11 At the face, Gary, Indiana (photograph from *Jacking Reinforced Concrete Pipe Lines*, Blue Book, by courtesy of American Concrete Pipe Association)

During the 1950s individuals and companies, quite independently, took up the pipejacking principle and developed their own equipment and methods for work in the UK, France, Germany and Scandinavia.

In Germany Ed Zublin first jacked concrete pipe in 1957 and promoted the technique both as manufacturer and contractor. Several other contractors also entered this field, so that by 1970 it was estimated that 200 km of pipe had been jacked in Germany [9].

In the UK, there are memories of cast iron pipe being jacked in the 1930s by a water company. The author reintroduced the use of pipejacking methods in 1958. The first job was to cross under the main rail lines at Glenton, north of Peterborough. Armco corrugated steel pipe was jacked as a sleeve over about 30 m (100 ft). Figures 1.12 and 1.13 show this job. Subsequently numerous steel casings were installed under rail tracks and highways. As other companies entered the field during the 1960s, competition increased and techniques advanced [10] [11].

The first work in Japan was in 1948 when a 600 mm cast iron pipe was installed under a railway using manual jacks. For some 10 years the main work was jacking cast iron or steel casings at crossings. Concrete pipes were first jacked towards the end of the 1950s [12].

During the 1960s and 1970s, techniques were refined to form the basis of present pipejacking methods. The most important advances were concrete pipes with rubber ring joints specially designed for jacking, shields with independent jacks to give steering corrections and the intermediate jacking station [11] [13].

Reasons for recent developments in pipejacking. The renewed interest in pipejacking was a natural consequence of the needs of the market. Throughout the Western World a massive replacement programme was put under way to make good war-damaged networks and to meet the demands of rising living standards. Much of the pipeline infrastructure had to be installed under busy roads and railways and at depths where open-cut was impracticable.

Pipejacking offered a solution which allowed short crossings to be made in a way that was inherently safe as well as economic. The ability to tunnel in diameters as small as 900 mm was one advantage. Another was that operatives could be trained more quickly in the skills needed for pipejacking than, say, in the skills required to drive a timbered heading.

To cope with differing ground conditions and to meet the varying needs of authorities in different countries, new methods were devised and equipment improved. Longer, larger bores became possible. By the late 1970s pipejacking was no longer being confined to crossings; in Japan and West Germany it was also being applied to sewer installation, traditionally done by conventional tunnelling [14].

Installing sewer pipes meant pipejacking at greater depths and in unstable ground conditions. It prompted a demand for rapid mechanised excavation and spoil disposal. Not surprisingly, a growing market sprang up for specialist machinery of many kinds. Mechanical cutting equipment, remote-control TBMs, screw, slurry and conveyer systems were introduced. Hydraulic rams, power packs and control systems were greatly improved. Contractors soon had access to more versatile shields, together with compact and efficient jacking rigs and intermediate jacking stations necessary for long drives and much more difficult ground conditions.

Figure 1.12 At the face, Glenton, Peterborough, England, 1958 (Author's photograph)

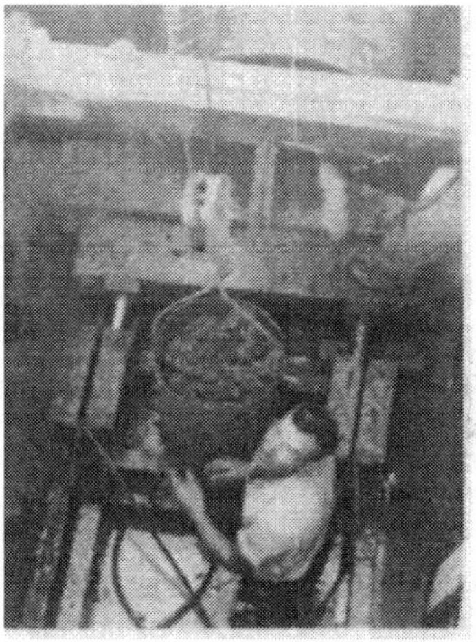

Figure 1.13 Jacking pit, Glenton, Peterborough, England (Author's photograph)

Not all this equipment was completely new. Mechanised cutting in the form of drum diggers and boom cutters had been introduced in the 1960s as had improved monitoring and guidance technology. The Japanese had introduced the first full-face slurry machine to pipejacking in 1965.

However, recent development has been so rapid that in many countries pipejacking is now the most prevalent form of tunnelled sewer installation for pipes and ducts up to 3000 mm, displacing segment lining and other traditional methods. Wherever there is a reasonably developed construction industry, there is now a contractor with pipejacking equipment and experience. On the other hand, for the highly sophisticated pipejacking systems we see more specialist contractors coming on the scene.

1.3.2 Microtunnelling

Microtunnelling develops in Japan: The production of microtunnellers was also market-driven. The Japanese, for whom open sewers had always been a traditional part of life, faced up to the enormous task of installing a new underground sewer network in extremely congested urban and suburban areas. The first stage of the programme was to install the larger main lines. This stimulated manufacturers to produce and contractors to use sophisticated slurry shields and other remote-control systems. As the installation programme progressed, the need shifted to smaller diameter installations. The industry responded with the microtunneller.

Komatsu appear to have started developing the first microtunnelling machines in 1972 and to have done their first job in 1974 [12]. These were of the pilot-bore type. Almost unknown outside Japan, this type of machine was installing about 30 km a year by 1979. The success of the pilot bore microtunnellers encouraged the manufacturers of the larger remote-control machines to progressively downsize their equipment, first to smaller pipejacking diameters and then to microtunnellers. The first pressurised slurry microtunnellers came on to the market around 1979.

Despite recent growth in other markets, the number of rigs operating in Japan is still several times greater than the total in the rest of the world. It is estimated that 400 contractors are operating between 1500 to 2000 microtunnellers.

Microtunnelling outside Japan. In the early 1980s, the West German Ministry of Research and Development funded a research project into ways of improving techniques for sewer installation. A study group visited Japan to find out more about the techniques being used there. Subsequently a Japanese Iseki 600 mm machine was imported into Germany for use on the Hamburg Microtunnelling Programme between 1981 and 1984, where 2500 m of small-diameter sewer pipe were installed [15]. The Hamburg project encouraged a number of German manufacturers and contractors to produce their own machines and also to develop smaller ones for installing house connections. The Germans remain the biggest users and the most important manufacturers of microtunnelling equipment in Europe.

The first microtunnelling installation in the UK came in 1984, but the method took about another six years to become established. Microtunnelling has since spread to all the countries of Western Europe. By the end of 1992 there were more than 200 microtunnelling machines in Europe, about 75% of these in Germany.

The introduction of microtunnnelling equipment into the USA had a false start in the early 1980s. Very little work was undertaken until a Houston contractor, BRH Garver, undertook a whole series of sewer contracts, starting in 1987, using Iseki and Soltau machines. Microtunnelling has been slowly gaining acceptance in North America since 1990. Microtunnelling methods are employed extensively in Singapore, Taiwan, Korea, Hong Kong and in Kuwait, Dubai and Abu Dhabi. Worldwide, there are now about 40 manufacturers offering microtunnellers ranging from simple house connection machines to sophisticated pressure balance slurry machines capable of drive lengths >200 m.

2 Pipejacking and microtunnelling methods

In its narrowest definition, pipejacking is a method of installing a lining. All the other elements — the shield, excavation and muck conveying — are much the same for other methods of tunnelling. In practice, it is the integration into a balanced system of all the activities of face control, excavation, spoil removal and lining that has made pipejacking a leading method of installing all types of pipe.

A pipejacking, or microtunnelling, operation consists of four elements — the face, the line, the jacking pit and the top side. Theses are shown in Figure 2.1.

2.1 Elements of pipejacking

2.1.1 The face

The activities at the face area are for:

- Excavating and controlling the ground.
- Loading the soil for conveyance.
- Monitoring line and level.
- Adjusting line and level.

These functions take place within some kind of shield fitted to the leading end of the pipe being jacked. The shield can range from a rudimentary steel cutting ring fixed to the leading pipe to a sophisticated earth pressure slurry shield with automatic computerised monitoring and guidance systems. Figure 2.2 illustrates some of the shield types in use.
The shield provides:

- A hard cutting edge to the net size required.
- A safe place from which to work.
- A mounting for mechanical excavation equipment.
- A mounting for face stabilisation devices.
- A place for monitoring line and level.
- A means of adjusting directional attitude to provide line and level correction.

Figure 2.1 Elements of pipejacking (drawing by courtesy of the Pipe Jacking Association, *A Guide to Pipe Jacking and Microtunnelling.)*

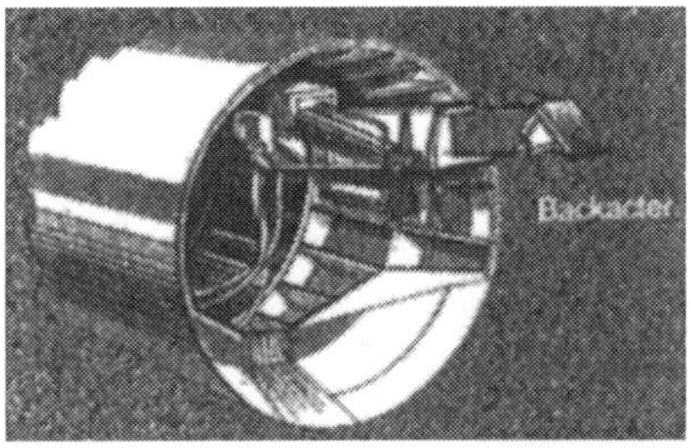

(a)

(b)

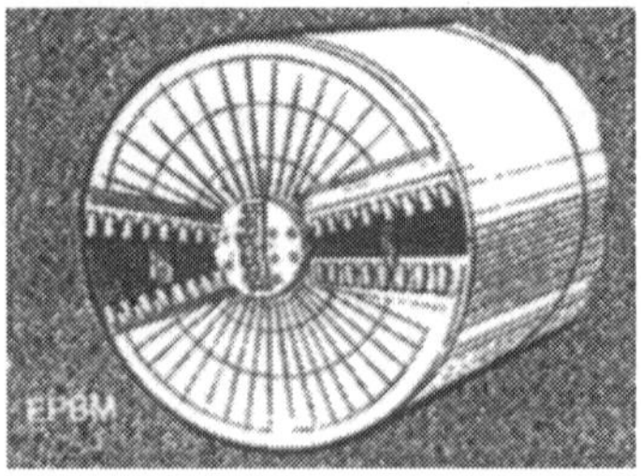

(c)

Figure 2.2 (a) Open shield with backacter, (b) Open shield with spoked cutter wheel, (c) Earth pressure balance shield. (drawings by courtesy of the Pipe Jacking Association)

In pipejacking, the shield may allow man-entry working and control, which is an effective approach, particularly for shorter drives in soils above the water table. For longer drives, mechanical cutting equipment is introduced into the shield. Where provision has to be made for unstable soils or working below the water table, remote-control pressure balance techniques are used.

Microtunnelling shields are miniaturised remotely-controlled versions of mechanised pipejacking. All microtunnelling shields have monitoring and control equipment with, in some cases, computer-controlled automatic adjustment.

Various forms of pipejacking and microtunnelling shields and TBMs are reviewed in detail in chapters 3 and 5.

2.1.2 The line

The line has several functions both during the work and during the design life:

During the work

- It provides a lining to the tunnel.
- It transmits the jacking loads without damage.
- It provides a passage for workers to get to the shield.
- It provides a passage for muck conveyance from the face.
- It allows services to be led to the shield.

During the design life

- It provides a permanent lining designed to give strength.
- It serves as a watertight and durable pipe suited to its use and location.

Figure 2.3 shows the line and its main features. The pipes and casings, which are pushed in to form the lining, are the most important element of the line. There are three main ways of installing the line:

- In a single pass, where the pipe driven becomes the permanent lining.
- In a double pass, where a temporary casing is first installed and then jacked out by the permanent pipes.
- Within a casing system, where the permanent pipe is laid within the outer duct and usually the annular space is filled.

The double-pass system is used when there is concern that pipe installed by a single pass will not meet the specifications, for example, where the finished line was a telecommunication multi-duct way. Double-pass, or two-stage installation, is also used in some microtunnelling systems.

The casing system is mainly for crossings, particularly under railways. Casings are used for crossings partly because of tradition and partly as a safety factor. When pipejacking was much less sophisticated, a casing provided a degree of tolerance. The safety benefits, however, are sometimes outweighed by other problems.

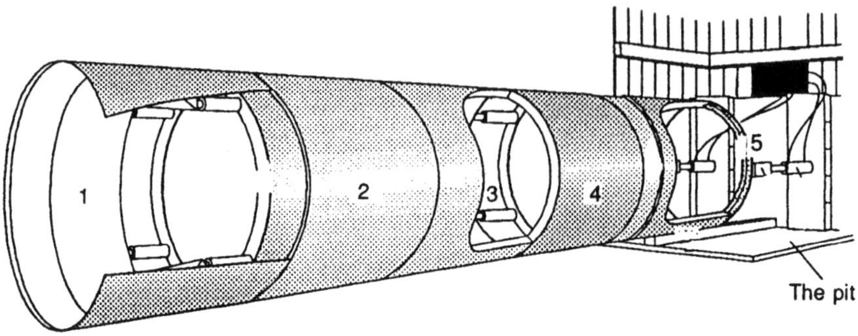

1 Jacking shield
2 Concrete lead pipe
3 Interjack station
4 Interjack pipe
5 Steel thrust ring

Figure 2.3 The line (adapted from a drawing by courtesy of the Pipe Jacking Association)

Types of pipe. Pipes used for jacking and microtunnelling include various forms of concrete, clay, plastic, steel, and ductile iron. Concrete forms predominate. Pipes have to provide adequate strength and permanence for their intended use and location. They also have to be strong enough to take the end loads, including eccentric loads, incurred during the jacking operation.

Pipe joints have to be watertight and withstand design test pressures. The joint has to be formed within the wall thickness to eliminate any external or internal projections. Some years ago, steel tubes with welded joints were popular as sleeves in which to carry a pipe (or pipes). Concrete is now the most commonly used material for sleeves. Its lower cost and lack of susceptibility to corrosion are advantages.

For gravity or non-pressure applications the most common form of joint is butt-faced with an external collar set within the exterior of the pipe diameter to provide a flush surface. Usually some form of sealing ring (or rings) is incorporated. The collar itself can be made of mild steel, stainless steel, GRP, or be of composite construction, as shown in. Figure 2.4. Other joint forms for concrete pipes include a version of the standard spigot-and-socket type, which is formed within the thickness of the pipe wall. An example is shown in Figure 2.5. Both types of joint illustrated have a load-distributing pipe packer located between the pipe faces.

For pressure pipes, steel mains have been used and these have been protected by a variety of materials. Joints are of the in-situ welded type. Ductile iron pipes with a specially developed bolted configuration and concrete sheathing are available in Europe and Japan. Concrete pressure pipes (prestressed and cylinder types) with modified joints have been successfully jacked. Pipes and joints are discussed in detail in Chapter 8.

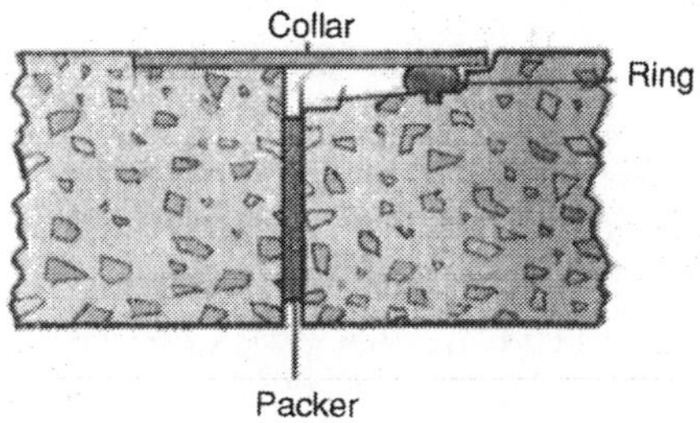

Figure 2.4 Sleeve joint
(Drawings by courtesy of the Pipe Jacking Association)

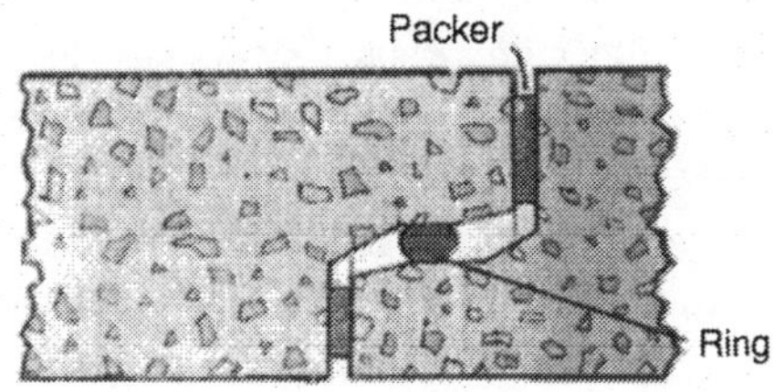

Figure 2.5 In-wall joint

Lengths of pipe sections. Pipe section lengths are determined by a number of factors, some of them conflicting. On the one hand the risk of leakage can be lessened by reducing the number of joints to a minimum, On the other hand longer pipe sections are heavier to handle and increase the length of the drive shaft; which can add significantly to job costs. Pipes are available in a range of lengths from 1 m to 5 m. In theory, no upper limit for diameters exists; pipes of DN 5000 mm have been driven. Box sections of much greater dimensions have also been installed to form road underpasses.

Drive lengths. The length of drive that can be successfully installed depends on the capacity of the jacking and the safe load that can be applied to the pipes. The jacking force required is partly determined by the weight of the pipes, but even more importantly by the surface friction between pipe and soil.

There are two ways of maximising the length of drive. The first is to inject suitable lubricants between pipe and soil, which can greatly reduce the skin friction. The second is to install Intermediate Jacking Stations (IJSs) in the line of the pipe. The IJS consists of a sliding steel can, which is to the same external diameter of the pipes and is fitted with a set of hydraulic rams. By operating these rams and using the pipes behind the IJS for reaction, the front pipes are pushed forward. The next operation is to retract these intermediate rams and push up the rear section into the gap created. By installing IJSs the load required to jack the total length can be divided between a number of jacking points. This reduces to acceptable levels the maximum load at any point. Figure 2.6 illustrates an IJS.

Single-span pipejacked drives of over 1000 m have been achieved with these methods. The length of drive that can be undertaken using inter-jack stations is theoretically unlimited, but practical considerations enter into the equation.

There is an important differentiation between the achievable length of drives in pipejacking and microtunnelling. For diameters below about 750 mm it is not physically possible to use IJSs as there is no means of accessing the line on completion to remove the hydraulic equipment and close the gap. In microtunnelling, the safe length that can be driven depends on the safe working load that the pipe can carry. The larger the diameter and the greater the wall thickness of the pipe the greater its load-bearing capability and the longer the lengths that can be safely jacked. Chapters 8 and 10 discuss in greater detail the design of pipes for permanent and temporary loads.

Muck disposal and links to the shield. Disposing of spoil and linking services to the shield are other important functions of the line. The speed and economics of a jacking operation are most often controlled by the rate at which spoil can be transported from the face and away. Depending on the size and nature of the work, a variety of methods can be used. These range from simple mine carts through mechanical systems like conveyors and augers to pumped slurry and pneumatic displacement systems. The line is also a means of providing power and communication links to the face and the shield for operation, monitoring and adjustment — all during the installation. Chapter 4 looks in greater detail at spoil removal methods.

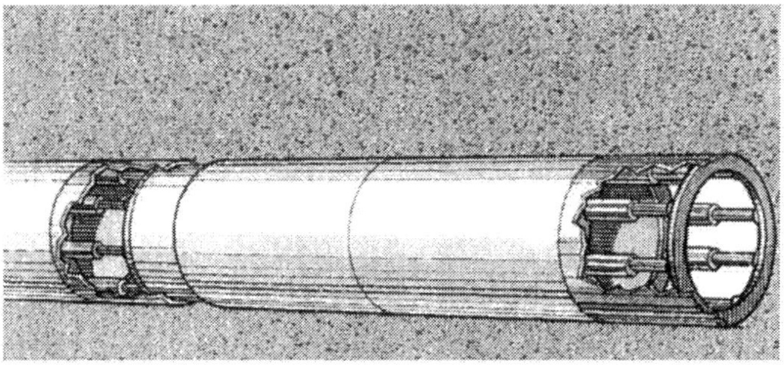

Figure 2.6 Intermediate jacking station (drawing by courtesy of the Pipe Jacking Association)

2.1.3 *The jacking pit*

The jacking pit must provide the following functions:

- Accommodate jacking equipment for pushing the line of pipes.
- Transfer the thrust uniformly on to the end of the pipe.
- Serve as a reaction element for the thrust.
- Provide an area for lowering, setting and jointing the pipes.
- Provide an area for spoil removal.
- Be a base for observation and a safe working place.

Fig 2.7 shows a typical layout for a drive pit.

Jacking rig. The provision of jacking capability is dependent on the safe load that can be applied to the pipe being jacked, which will in turn depend on pipe diameter and wall thickness. This safe load can vary from less than 50 tonnes in the case of a small microtunnelling job to several thousand tonnes in the case of a large pipe section. A generous reserve is normally provided to accommodate high short-term start-up loads that can occur after a shutdown of work. To provide the thrust, two or more hydraulic rams are normally mounted on a frame. These rams or jacks can be long stroke, whereby they are matched to the length of the pipe section to be driven through the pit, or short stroke, which requires spacers to be inserted to provide the travel through the pit. Alternatively, after each stroke the jacks are moved forward along the frame, then fixed and the jacking cycle repeated.

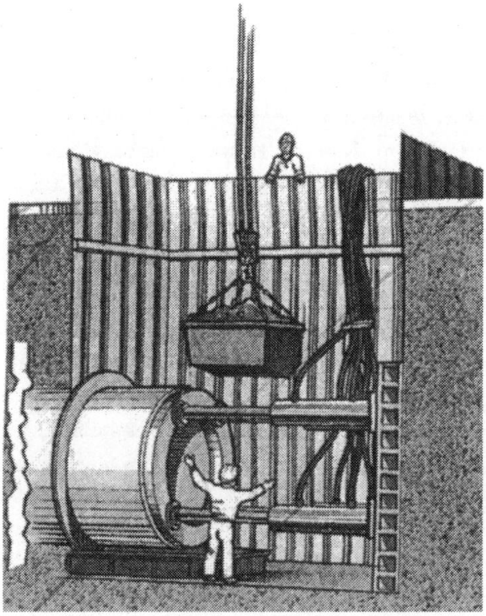

Figure 2.7 Drive pit and jacking rig (Drawing by courtesy of the Pipe Jacking Association)

Thrust ring. To transfer the thrust from the jacks uniformly on to the pipe-end, a jacking or thrust ring is provided. Generally of steel, this is designed to fit snugly to the pipe-end profile. The ring needs to have sufficient depth to spread the load uniformly. Chapters 4 and 5 contain additional details on jacking rigs and thrust rings.

Reaction wall. Behind the jacks it is necessary to provide a reaction element capable of taking the maximum anticipated jacking load without movement. Most commonly this is a wall at the back of the drive pit which distributes the loads from the rams safely into the undisturbed earth behind the pit. When working above ground, or where the soil conditions are not capable of absorbing the load, it may be necessary to provide alternative reaction by blocks, piles, ties or other means.

A common problem is the failure of the reaction element. Failure to understand how various soils behave under load sometimes leads to excessive movement of the reaction wall and ultimate failure. More information on the design of reaction walls can be found in Chapter 11.

Pipe installation. Successive pipes have to be lowered into the pit, set into the correct position and jointed. Where remote-control methods are being used, the spoil removal system, together with power and communication lines, have to be disconnected as each pipe reaches the end of the pit. They are then reconnected after the new pipe is in place. The pipe rests on a supporting cradle as it is jacked through the pit.

In non-cohesive soils or where working is below the water table, a sealing *eye* must be formed in the shaft wall to stop soil and water flowing around the pipe and into the shaft.

Monitoring. The pit provides the base for observation and monitoring of the line and level. A simple method is to set up a survey instrument, such as a theodolite, on a known base station to observe a target at the face. More commonly a laser is fixed in the shaft at the known base to provide a constant beam of light focused on the target. The monitoring and guidance of shields is considered in detail in Chapter 3.

Shaft size and design. The size and design of the jacking and reception shaft is determined by a number of factors: the diameter and length of the pipe; the type of shield and the size of its largest unit; the jacking cylinder and rig configuration.

The cost of the drive shaft, particularly in deep sewer work, is a significant part of the overall cost of the job. In both pipejacking and microtunnelling, the trend is to produce equipment of minimal dimension that can operate from a circular shaft. These can be readily adapted to form the permanent access chamber. Closer consideration of shaft dimension and design is given in Chapter 11.

2.1.4 Top side - surface equipment

The area round the top of the shaft is organised to provide a number of support functions and services, which include:

- Muck handling, cleaning, storage and disposal.
- Primary power units - electric, hydraulic and pneumatic.
- Lubrication-mixing and injection equipment.
- Pipe handling.
- Pipe storage.
- The control station.

Muck and pipe handling. Methods of raising the muck from the bottom of the shaft to the surface vary from simple skips hoisted by a crane to pressure pipelines lifting the spoil and discharging it as slurry into storage or separation containers. Where slurry is pumped, the surface equipment will include a means of separating the bentonite slurry from the muck. This can be by settlement, centrifuges or other devices. The muck is transferred to hoppers for disposal and the bentonite returned to the slurry system. Spoil handling is considered more fully in Chapter 5.

Pipe sections are normally dropped down the pit by some form of cranage.

Primary units. Primary power units provide electric and hydraulic power to operate the shield motors, hydraulic systems, muck disposal methods, hoists and other equipment. An air compressor may be needed to provide pneumatic power for cutting tools, muck disposal systems, air locks and shields. The equipment used for mixing and injecting the lubricating fluids is also located close to the head of the shaft.

Where space is not at a premium, or in a badly designed or ill-organised installation, the surface area taken up by all this equipment can be very extensive. The need for the smallest site *footprint* to reduce interference and minimise disruption has created a demand for compact working. Pipejacking and microtunnelling equipment is being produced on an integrated basis where all the elements are part of a system. Increasingly common is a modified standard container which incorporates cranage, power, lubrication equipment, monitoring and control.

With microtunnelling and small diameter remote pipejacking, the control centre is often located at the surface, with all monitoring information and control commands under the control of the operator — or sometimes a computer. With the rationalisation of pipe and muck storage, it is feasible to accommodate a complete microtunnelling set-up in a surface area of 20 m by 4 m, or less where space is particularly restricted. Figures 2.8 and 2.9 illustrate different surface installations.

2.2 Applications of pipejacking and microtunnelling

There are four main applications for pipejacking and microtunnelling:

(i) Line installation	(ii) Installation of ducts
- gravity sewers	(iii) Crossings
- pressure pipes	(iv) Pipe replacement

2.2.1 Line Installation

Sewers. The greatest volume of work for pipejacking and microtunnelling is for the direct installation of new gravity sewer lines. Techniques for both have been developed to meet the specific characteristics of stormwater and sanitary sewer installation.

Figure 2.8 Large pipejacking - surface installation (photograph by courtesy of Euro-Iseki Ltd)

Figure 2.9 Microtunnelling - surface installation (photograph by courtesy of Dr. Ing. G. Soltau GmbH)

Characteristics of stormwater and sanitary sewer installation are:

- Installation has to be to close line and level tolerance — typically ±25 mm.
- A significant volume of sewer installation is for pipe of DN 300-2000 mm.
- Gravity flow requires pipes to be laid progressively deeper and with larger diameters.
- Sewers, wherever possible, are laid in straight lines.
- Sewers are predominantly located in the highway.
- Access shafts have to be provided at changes of direction and main pipe intersections.
- Access shafts for non-entry pipes (<900 mm) are normally located no more than 120 m apart for purposes of maintenance.
- Man-entry-sized pipes often call for a greater distance between access shafts.
- Pipes have to be resistant to corrosion and abrasion, watertight and capable of withstanding live and dead loads.

Properly carried out, modern pipejacking and microtunnelling techniques can meet these criteria, and thereby the client's technical needs. Pipejacking is used for the installation of pipes of DN 900 mm or greater. There is no theoretical upper limit on diameter, but in practice most installation is for sizes up to 2500 mm. Pipejacking is not limited to installation of circular sections; a variety of rectangular box culverts have been driven. Microtunnelling is used for installing sewer pipes of DN 250 mm to less than 900 mm. Normally only circular sections can be driven, although circular pipes with an egg-shaped interior have been installed.

For most sewer work, clients' needs on length between shafts can be directly met without using temporary intermediate shafts. Broadly, as the diameter increases so does the possible length of drive between shafts. The maximum load that can be applied to a 300 mm pipe may limit the drive length to less than 100 m, whereas a 600 mm pipe can be driven 150 m safely. At 1000 mm the drive could be much longer, but safety concerns will probably limit the length to 200 m. This length will rise to 600 m drives at DN 1500 mm, and to 1000 m at DN 2000 mm.

The installation of sewers by pipejacking and microtunnelling can compete directly on price with the alternatives of deep open cut and traditional tunnelling. Such methods can offer a viable alternative at depths greater than 4 m or where the installation is under busy roads or close to exiting buries service lines. These methods will also be cost-effective if ground conditions are unstable, where work is below the water table or there is a risk of settlement. In environmentally sensitive areas or where the works will cause disruption to the community or be detrimental to local business, the social costs incurred by open-trench methods of installation will make pipejacking and microtunnelling competitive alternatives.

In Japan and Europe, pipejacking has replaced traditional tunnelling as the predominant method of installing underground sewers in DN 900-2500 mm. Microtunnelling, with its later introduction in the late 1970s, is now rapidly gaining acceptance, and will continue to command an increasing share of sewer work.

According to figures published at the Osaka No-Dig Conference [12], it was estimated that about 10% of the 14 000 km of pipeline installed in Japan in 1990 would be by some form of jacking, including microtunnelling. This percentage had risen steadily from 5.6% in 1980. The rapid penetration of the market is even more remarkable in view of the fact that in 1960 virtually no sewers were directly installed by these methods. Figure 2.10 shows an interesting application of direct installation where a sewer has been laid on a curved path.

House connections. Non-disruptive ways of installing lateral connections to properties are now also in demand. As a consequence, systems have been produced for installing sewer pipes in DN 150-250 mm to the required line and level. This equipment is designed to work from restricted areas, including basements and small inspection chambers, with drive lengths typically up to 30 m. Figure 2.11 shows a microtunnelling system designed to install house connections.

Pressure pipes. The characteristics of pressure line installation generally favour jacking methods less. For pressure pipes there is no requirement for close line and level tolerance and seldom a need for deep installation. Despite this, there is a small but increasing demand for direct installation of pressure pipes, either as an alternative to tunnelling or in circumstances which make open cut unattractive. The main demand comes for installation of water transmission lines. The development of pipes and joints suitable for jacking but which also meet the needs of pressure installation has made this possible. In Chapter 8 we look in detail at these types of pipe. Up till now the greatest volume of pressure pipe installation has been in Japan.

2.2.2 Installation of ducts

Permanent ducts are required for a variety of purposes, for instance, laying cable and pipe tunnels when permanent access is needed. Pipejacking and microtunnelling are satisfactory solutions where ducts have to be laid deeper or in a non-disruptive manner.

A large diameter pipe with a watertight joint can be jacked in to provide a suitable duct for a range of small pipes and cables. Figure 2.12 shows a concrete pipe carrying service lines and providing man-entry. Ducts of this type are in increasing demand for carrying power transmission cables in suburban and urban areas, as overhead cabling, the traditional alternative, is no longer acceptable in many cities. Multi-way ducts, popular for power and telecommunication cables, provide other installation opportunities for pipejacking and microtunnelling. With a two-pass system, it is possible to install precast sections with multi-duct ways already formed, as shown in Figure 2.13.

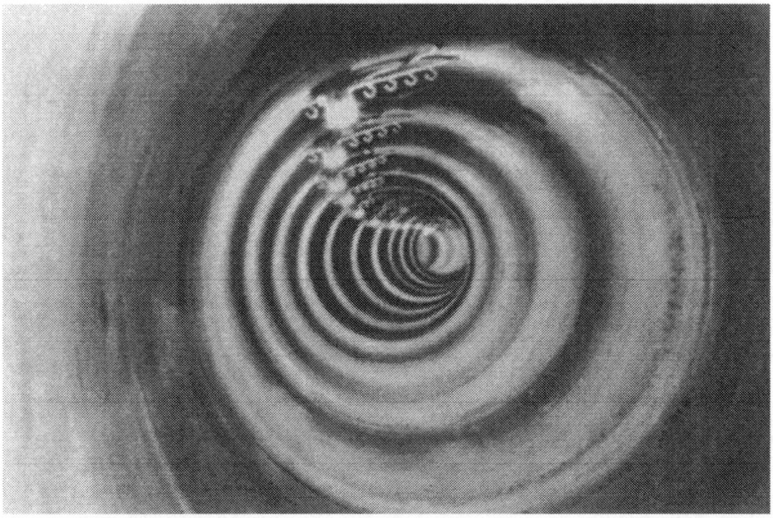

Figure 2.10 A curved pipejacked installation

Figure 2.11 Soltau RVS 35 house connection system (photograph by courtesy of Dr. Ing. G. Soltau GmbH)

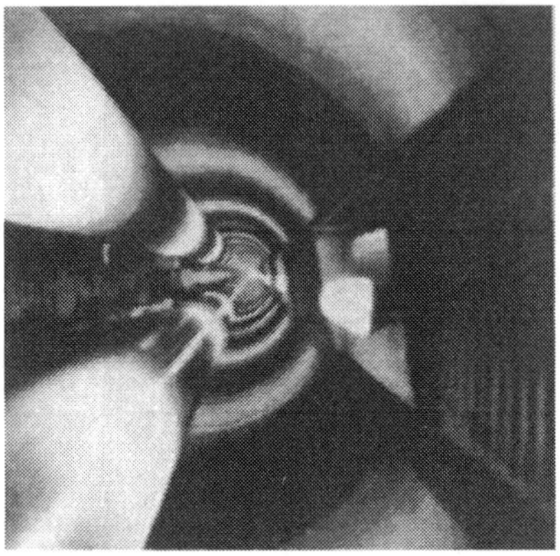

Figure 2.12 Concrete pipe carrying lines with man entry

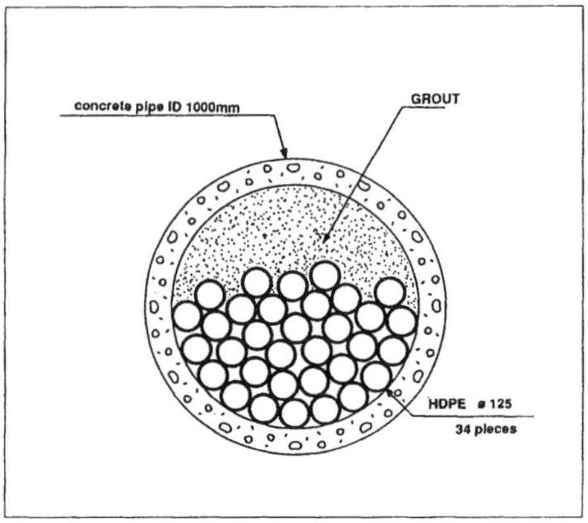

Figure 2.13 Concrete pipe with multi ducts (drawing by courtesy of Smet-Boring)

2.2.3 *Crossings*

Installing a casing at a highway or railway crossing is the traditional use of pipejacking. Within this casing the operational pipeline is installed and the annular space normally filled. The casing, often a steel tube with welded joints, is in reality a temporary liner to the bore.

More than 60 years ago it became apparent that pipejacking offered the possibility to direct-install concrete pipes at crossings to provide a permanent liner, not only for sewers and culverts, but also for larger structures, such as pedestrian underpasses. This underpass application has been further extended by the use of box sections. Full-sized vehicle underpasses are installed by jacking insitu-cast sections under existing highways and railways.

The crossing market continues to be important for both the installation of casings and for the direct installation of pipes and sections. All pipe and cable utilities generate pipejacking work where their lines cross railways, highways, waterways and other obstructions, and also where economics or legislation require a non-disruptive installation method. Probably the greatest demand comes from cross-country transmission pipelines for gas, water and oil. Several thousand individual crossings are installed every year by pipejacking.

Installing casings. The requirement that a casing be installed arises from legislation framed by railway and highway authorities some years ago. The need for a casing is being increasingly questioned and direct installation of a pipe is now generally gaining acceptance — except in North America.

Casing work is characterised by relatively short drives rarely exceeding 100 m. Casings are often installed at relatively shallow depths and work is therefore usually in stable soil conditions and above the water table. Line and level tolerance requirements are likely to be less severe than for direct installation. Such working conditions suit open shields with hand or mechanical excavation rather than full-face remote-control machines. For casing diameters less than 900 mm, microtunnelling has found a limited use. The alternative of auger boring offers a cheaper alternative where ground conditions are suitable (see Chapter 14).

A major advantage of a properly designed and controlled pipejacking or microtunnelling crossing operation is that work can be carried out with minimum soil cover (1 m or less) and eliminate or reduce soil movement and subsidence to a tolerable level. Figure 2.14 shows an example of a cased crossing installation.

Figure 2.14 A 750 mm cased installation - Honolulu International Airport (photograph by courtesy of Herrenknecht Corp.)

2.2.4 Pipe excavation techniques

The on-line replacement of sewer pipes by pipejacking and microtunnelling is a relatively new development. So far this represents only a small amount of work, mainly in Japan and Germany, although some on-line replacement has also been done in other countries.

A number of remote-control microtunnelling and pipejacking TBMs have rock and boulder cutting capability. This capability has now been extended and applied to replacing pipes by excavation. As the old pipe is broken out and removed, the replacement pipe is installed behind the TBM, in the way common to most trenchless pipe-replacement techniques. Figure 2.15 illustrates the concept of pipe excavation — often termed *pipe eating*.

Pipe excavation equipment is capable of :

- Removing the old pipe and installing the new.
- Providing a pipe of equal or larger diameter.
- Installing to a revised or corrected invert level.

Some excavation systems have been developed with the additional capabilities of:

- Removing old reinforced concrete pipes.
- Maintaining sewer flow during replacement.

Figure 2.15 Pipe excavation — a pipe-eating mole (photograph by courtesy of NLW Fördertechnik)

The current range of pipe diameters that can be removed has been between 150 mm and 600 mm, although there seems to be no technical reason why larger diameter pipes could not also be replaced in this way. Replacement pipes with diameters up to 700 mm have been installed.

2.3 Advantages and limitations of pipejacking and microtunnelling

The advantages and limitations of using pipejacking or microtunnelling as a solution have to be related to the work in hand, its particular problems and to the benefits or otherwise of alternative methods. If we take clients' needs for typical sewer installations one by one, we can compare the advantages and limitations of pipejacking and microtunnelling for each of these with traditional construction methods of open-cut trenching and tunnelling.

Generally the client needs will be:

- To work in a wide range of ground and water table conditions .
- To cause minimal damage to property and existing utilities.
- To minimise disruption to the public and environment.
- To ensure safe conditions for workers and the public.
- To install to the owner's specification.
- To provide a cost-effective solution.

2.3.1 Working in a wide range of ground and water table conditions

Thorough advance site investigation and evaluation is essential before any rational judgement can be made about the best installation method to be used and the measures to be taken. This applies whether traditional or trenchless solutions are adopted. The importance of this first investigatory stage cannot be overemphasised.

Pipejacking and microtunnelling methods are now available that can successfully deal with most types of ground condition, from hard rock to highly unstable silts and sands with a high water table. Equipment and boring heads have been designed to deal not only with high water tables in non-cohesive soils, but also with difficult mixed ground, such as sands containing cobbles. Among the most troublesome ground conditions are clays with large boulders, of the type occurring in glaciated areas.

Microtunnelling has more limitations than pipejacking in terms of ground conditions. The smaller diameters and remote aspect of the operation make microtunnelling more prone to problems where soil conditions change rapidly or when obstructions are encountered.

As depth and difficulty increase, jacking methods have greater advantages over traditional open-cut construction. Trenching below the water table in difficult ground will require expensive provision for shoring, stabilising and dewatering.

Compared to tunnelling, on a size for size basis, there are no clear-cut advantages or limitations in dealing with ground conditions. The type of shields and techniques used for jacking and tunnelling are almost identical. One area where segment tunnelling could have an advantage is when work is in overconsolidated and expanding clays. In such conditions high frictional loads can develop which limit jacked drive lengths.

2.3.2 Minimising damage to property and existing utilities

An increasing amount of sewer construction is now to replace or reinforce the existing network and, by definition, takes place in developed areas where risks are greater of ground movements damaging property, pavements and other utilities. This is a growing concern for many owners.

Jacking methods are inherently some of the safest and least disturbing techniques of underground construction. Pipejacking and microtunnelling are designed to provide a continuous lining as the tunnel advances, without significant overcut of the soil. Ground movement is therefore greatly reduced or avoided altogether.

The external shield and pipe diameters should be very nearly the same, with a minimum of overbreak. For working below the water table and in difficult ground, earth pressure balance and slurry pressure shields are available which can be operated to ensure that neither ground heave nor settlement occurs. With care and a proper choice of technique, jacking methods should cause little ground disturbance. Below heavily trafficked roads and railways, sewers have been successfully installed with only 1 m of cover. Existing foundations or services have been passed within a few millimetres without any measurable movement.

In reducing damage and risk of settlement, jacking methods offer considerable advantages over open-cut working. Deep trenches in difficult ground will almost inevitably cause movements resulting from operations to lower ground water and to install and withdraw temporary shoring and backfilling.

In traditional tunnelling, the segmental lining is placed within the tail of the shield. This leaves an annular gap between the excavated tunnel diameter and the outside diameter of the lining. Where the ground is stable this void can be filled with stabilising grout. However, in unstable ground it is difficult, if not impossible, to avoid ground moving into the space before the grouting is done and thereby causing settlement.

2.3.3 Minimising disruption to the public and the environment

The objectives of early pipejacking applications were to install a pipe under a highway or railway without interrupting services. Traditional trenching methods of installing all forms of utility are now being seen as potentially damaging and disruptive if the work is in the road or in environmentally sensitive areas. Sewer construction is perhaps the most disruptive of all utility work for the simple reason that sewers are normally located in the roadway and have to be laid to straight lines and at greater depth. When sewers are constructed in a busy central street, the costs associated with traffic delays, loss of business and general disruption can be several times the value of the work itself. When a pipe has to be laid under an environmentally sensitive area, it is obviously better to pass under it rather than through it.

All underground construction methods, including both jacking and tunnelling, can reduce disruption by confining surface works to a number of shaft positions. If the location of shafts is carefully designed and planned, the impact of the works can be limited further. Present equipment and working methods not only require a smaller site area for shafts but are also quicker than open-cut methods of installation.

2.3.4 Safe working for operatives and public safety

Pipejacking operators work in greater safety than their counterparts in open trenching and tunnelling. Safely excavating deep trenches or driving segment tunnels also requires skill and experience. Jacking demands less in the way of experience and traditional skills but more in technical competence. Even when jacking is with an open shield, the man at the face is well-protected from cave-ins. When excavation is mechanised, the equipment is controlled by just one or two men from the rear of the shield. Microtunnelling operations, implicitly fully automated and remotely controlled, are increasingly being applied to larger diameters. No operatives enter the tunnel during driving; activities are confined to the drive shaft and the operator's panel is in a site cabin.

Public safety is greater where jacking methods are employed, as the impact and area of construction works are reduced. Site works are also easier to protect and control.

2.3.5 Installing to owner's specification

An owner is concerned more with the permanent works, the pipes and the access shafts, and less with the method of construction. His needs are the same whatever the method of installation and the objective must be to meet these. (Clients' needs in relation to sewers were set out in section 2.2.1). Pipejacking and microtunnelling have now developed to the extent that, in nearly all cases, they can meet these client needs for pipes, accuracies, drive spans and curves. It must be said, however, that in some countries this ability has still to be fully achieved and acceptance has still to be given.

Pipes. The pipe manufacturing industry has responded to changing demands by producing modified pipe designs suitable for pipejacking and microtunnelling. Owners can choose from pipe materials which include unreinforced, reinforced and prestressed concrete, clay, fibre-reinforced plastic mortar (FRPM), plastic, ductile iron and steel. Joints can meet both gravity and pressure pipe specifications. These installation methods demand high-quality heavy-wall pipe, manufactured to close tolerances. Such specifications mean the final product will be of a higher quality than the equivalent product installed by open trenching. Pipes available for jacking range from 150 mm to 5000 mm, covering most client needs.

The direct installation of factory-made pipes with traditional proven joints gives jacking methods a great advantage over tunnelling. Traditional tunnelling installs a segmental or other form of liner which is not suitable for use as a finished pipeline, therefore it is generally necessary to introduce a secondary liner pipe. The other drawback for tunnelling is that the minimum size of segment tunnel that can be installed is 1200 mm.

Accuracies. Both for microtunnelling and pipejacking it is possible to achieve the same kind of accuracies that are obtained in practice in trenching work. A generally accepted installation tolerance for microtunnelling and jacking is ±35 mm on the vertical and horizontal axis. The introduction of improved monitoring and guidance control allows even higher standards of accuracy.

Drive spans. There are limitations to the drive spans achievable by jacking between shafts. For pipejacking these limitations are more practical than technical, as spans of more than 1500 m have been driven with the use of IJSs. However, for reasons of safety and economics, as diameters drop from DN 2000 mm, drive spans should reduce accordingly. Although there are no rules, experience indicates that maximum spans of 500 m are sensible for DN 1500 mm pipe, and perhaps 350 m for DN 1200 mm.

For microtunnelling the limitations are more real. The restriction on drive span is imposed by the jacking load limit that can be applied to the pipe. So far it has not been practicable to introduce inter-jack stations into pipes of internal diameters less than 750 mm. The distance is also limited for various types of monitoring system. The larger the diameter, the thicker the pipe wall (and thus the cross-section area), and therefore the greater the safe loads that can be applied and the longer the spans that can be installed. As non-man-entry pipe sizes require closer shaft spacing for operational reasons, modern microtunnelling equipment and pipes are capable of meeting all but the most specialist needs. Drives of over 150 m are regularly achieved with pipe diameters of 600 mm and upwards.

Driving on curves. A basic limitation of pipejacking is that it is not possible to make a rapid change of line or level or introduce a bend into the line. This is a clear advantage for traditional methods of working. However, numerous jobs have involved driving all or part of the length to a slow radius curve. Vertical, horizontal and reverse curves have been driven. The radius that can be safely driven depends on the pipe joint configuration and soil conditions but is often limited to about 100 m in Europe.

2.3.6 Providing a cost-effective solution

At shallow depths and where the costs of reinstatement or disruption are minimal, open-cut trenching remains the economic choice. As various factors come into play, singly or in combination, costs of open-cut escalate rapidly, at which point pipejacking and microtunnelling become more cost-effective. The economic factors influencing the decision to select a jacking solution are discussed in detail in Chapter 14.

Undoubtedly all three methods — open trenching, tunnelling and jacking — have their role to play in utility installation programmes. What has become apparent in the last two decades is that microtunnelling and pipejacking offer technical and economic advantages for an increasing number of situations, and particularly in sewer construction. It is therefore not surprising that these trenchless methods have already achieved significant market penetration at the expense of the more traditional approach.

3 Equipment: shields and tunnel boring machines

3.1 Introduction

The work undertaken at the face is critical to the jacking operation. The maintenance of ground stability, the speed of working, control of line and level and the safety of operatives are all related to the choice of shield or tunnel boring machine (TBM). Choosing the right machine to suit ground and water table conditions is discussed more fully in Chapter 9.

Shields and TBMs for jacking are similar in principle and operation to those employed in traditional forms of tunnelling. The most rudimentary form of shield, of the type used in early jacking, is a steel cutting cylinder, without provision for steering adjustment, fitted to the leading pipe. This cutting edge, often permanently fixed to the first pipe, was not recovered on completion of the drive. Cutting rings are rarely used today and only then for short drives in stable soils.

Whatever the type of pipejacking shield, the tail of the shield slides onto a recess formed on the outer diameter of the first pipe or section being pushed up behind. When working below the water table or where ingress of fine material needs to be prevented, a tail seal is provided between shield and pipe. A spreader ring transfers the loads between shield and pipe. The difference between the outer diameters of shield and pipe, the *overcut*, is normally kept to a minimum.

Almost all shields used today are provided with some form of steering adjustment which allows the shield to be moved independently of the pipe being jacked. For the most part this is to provide angular adjustment, but in some shields the hydraulic jacks can be used to move the shield forward independently of the lead pipe. The larger and more sophisticated shields are of articulated construction, with a flexible joint between leading and tail sections to allow for more flexible operation and better control.

3.2 Classes of equipment

Steerable shields can be classified into four main groups: conventional, compressed air, auger and pressure chamber shields.

3.2.1 Conventional (open) shields

Conventional shields, often called *open shields*, are defined as those which allow direct access to the face, or to a point immediately behind the face cutters. Relatively simple, flexible and cheap, these shields are used for shorter drives in stable ground conditions. However various modifications have been introduced to control the face stability and speed up excavation and spoil disposal.

Conventional shields can again be classified into manual, blind, semi-mechanical and mechanical, with a number of variations within each of these classes, as listed in Table 3.1.

Table 3.1 Conventional shields

Type	Variations
Open manual (Excavation by hand)	Compartments Bench plates Spiling Breast plates Doors
Blind	Dome/disc
Semi-mechanical	Modified back hoe Boom cutter Telescopic arm - bucket/spade - rotary cutter
Mechanical (rotary)	Spoke cutter arms Disc Dome

Open manual shields. A simple manual shield for use with a concrete pipe is shown in Figure 3.1. Pipes in diameters ranging from 900 mm to 5000 mm and non-circular sections, such as horseshoe and rectangular boxes, are installed using shields of this type. Figure 3.2 shows a shield for box jacking. Manual shields would normally be employed for drives of less than 200 m.

To improve the stability of the face, a number of modifications can be made:

- The profile of the end of the shield can be inclined from a nearly vertical 80° for stiff clays to 60° for sands and gravels. The greater the inclination the more difficult it is to steer a shield.
- The shield may be fitted with diaphragm stiffeners, which compartmentalise the face and allow individual areas to be worked. This technique is more often used for larger diameters.

- Horizontal bench plates can be fitted, which provide stabilising planes for sandy material.
- The traditional heading technique of *forepoling* can be mimicked by having *poling plates* or *spiles* set in the top of the shield. These can be independently pushed forward to provide top support.
- Shields can be fitted with retractable breast plates to hold up portions of the face.
- Doors can be fitted which allow the whole face to be closed off.

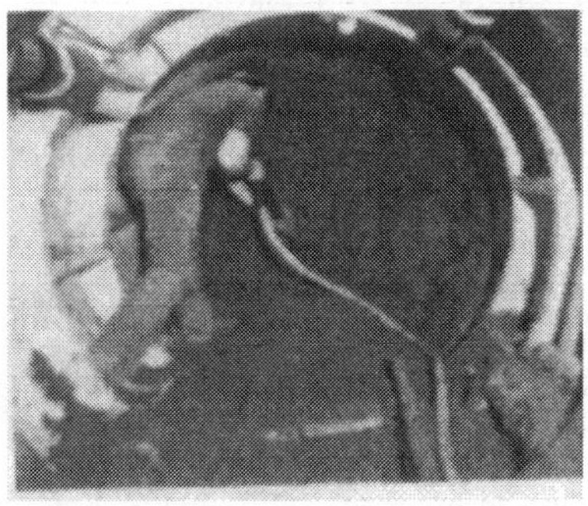

Figure 3.1 Manual pipejacking shields (photograph and drawing by courtesy of the Pipe Jacking Association)

Figure 3.2 Box shield for pipe jacking (photograph by courtesy of Laserbore)

Blind shields. When jacking in soft homogeneous plastic soils (N values less than 3) virtually the whole face can be blanked off by a disc or dome. The soil is forced through apertures, or slits, which are between 2% and 10% of the face area. The soil, which needs to have low permeability and cohesion, is remoulded as it is squeezed through the apertures. This allows some control on the flow of the material. Care needs to be taken to ensure that over-pressurising the soil does not cause outward soil movement and heave. Such shields are forerunners of *earth pressure balance shields.*

An interesting example of this type of blind shield is shown in Figure 3.3. A shield of extremely light weight construction is provided with a central port with four quadrant doors. Any one or combination of doors can be opened by the operative, allowing the soil to be forced through the opening in the form and consistency of toothpaste. In very soft soils (N values of 1 or less) it is possible to steer the shield by differential opening and closing of the quadrants.

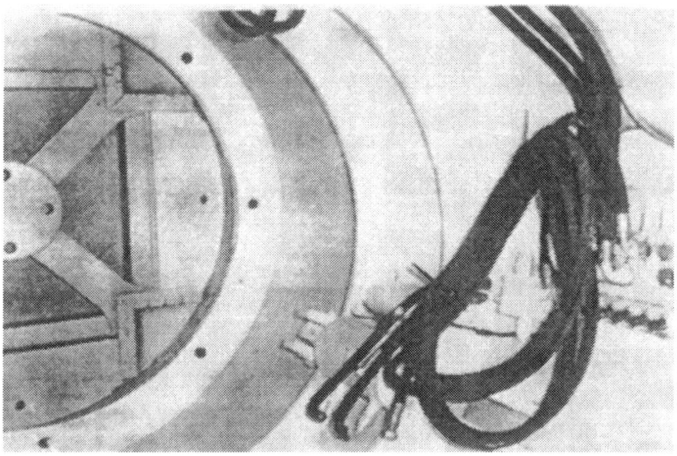

Figure 3.3 Blind shield with quadrant doors (photograph by courtesy of Decon Engineering Company)

The Swedish Lundby system, which lays pipes in the diameter range 800—2700 mm, is designed for soft ground and has no excavating equipment. It has a blind shield with an earth pressure balance controllable by a hydraulic door at the rear of the leading section. The shield comprises a cutting ring with a hollow centre. Soil is forced into the head through the doors and wire ropes cut the material into smaller pieces. The spoil builds up until the laser beam is obscured, at which point a scraper pulls the spoil back to the shaft. Figure 3.4 shows a 1400 mm Lundby shield with the control door.

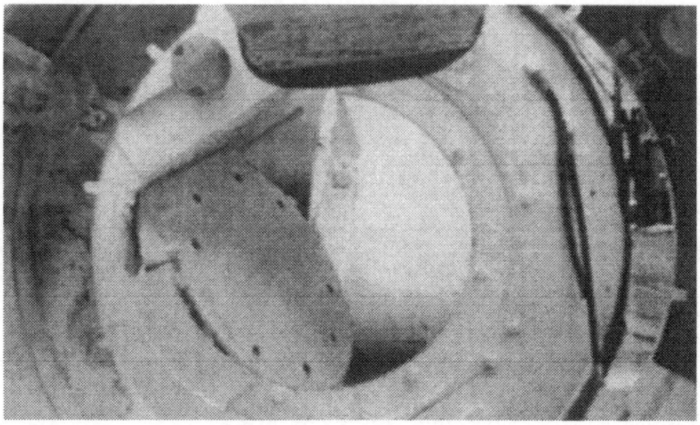

Figure 3.4 Lundby system 1300 mm shield with control door (illustration by courtesy of Lundby System AB)

Semi-mechanical shields. Adding some form of powered excavation device can improve production. The operator of a semi-mechanical shield excavates the face by manipulating a cutting head mounted on an arm round the face. To match the greater output of this type of equipment, a conveyor, or screw auger, disposes of the muck from the face. Some excavators are fixed and mounted in the roof of the shield to leave the invert free. Others, floor-mounted or independent, can be run in and out of the tunnel. Variations on such equipment include:

- Modified hydraulic back-acter arms with buckets or spade cutters.
- Boom cutter arms with rotary cutting heads suited to harder soil conditions.
- Telescopic arms which can be fitted with hydraulically-operated bucket, spade cutters or a rotating cutting-head.

Mechanical shields with rotary cutting heads. Rotary cutting-heads are driven from a central shaft or annular-mounted motors. The rotating arm or plate is fitted with cutters which sweep and cut the whole face. Face arrangements can vary from a single beam with two spokes mounted with cutters on the face, through three-, four-, five- and six-spoke versions.

For less stable ground conditions, the spokes are replaced by a plate or dome-type full face which has apertures or slits to allow the soil to pass into the shield. A full-face cutterhead of this type can provide considerable support and stability to the face. This stabilising effect can be greatly enhanced by independent movement of the cutter-head within the shield. Retracting the head into the shield mobilises an earth pressure balance effect. The volume and speed of excavation with a full-face machine require mechanical systems for disposing of the spoil. Figures 3.5 and 3.6 show a simple spoke shield driven from a central shaft and a full face rotary machine respectively. Rotary machines of this type are used on drive lengths up to 500 m, although longer lengths may be possible in some ground conditions.

Figure 3.5 Open shield with rotary cutter bar (photograph by courtesy of Akkerman Co. Inc.)

Figure 3.6 Full face rotary shield (illustration by courtesy of Komatsu, Japan)

Conventional shields in general. All the above techniques, classified as conventional shields, are suitable for a wide range of cohesive soils including rock, clays and silty clays, which can be below the ground water table. Non-cohesive soils above the water table are also suitable, but become less viable with ground water pressure. A great advantage of open shields with access to the face is that they allow operatives to identify and deal with any natural or artificial obstructions.

Rotary shields of all types are often classified under the general heading of tunnel boring machines (TBMs).

3.2.2 Compressed-air shields

Compressed-air working, frequently used in tunnelling as a means of applying a counter-balancing pressure to hold back soil and ground water, has been adapted to jacking. The most common method takes advantage of the moving line of pipes and limits the locked most common method takes advantage of the moving line of pipes and limits the locked section to the front part of the drive only. In this way only the face, the shield and first two or three pipes are under compressed air; the rest of the line and the shaft remain in free air. The lock automatically travels forward as the pipes are jacked in. In the right conditions a travelling lock can provide an effective solution. However it has to be recognised that the reservoir of air with a travelling lock is relatively small and any sudden loss of air will dramatically change the operating pressure.

Compressed air can also be operated in a manner similar to that used in tunnelling. By setting the lock at the head of the shaft, all work in the line, the shield and the shaft bottom is under compressed air. To avoid men working in compressed air, the head only can be locked off, leaving the operator in free air to handle the excavating equipment by observation through a glass panel or with the aid of a cctv camera. It is still possible for a man to enter the chamber if necessary — to remove boulders, for example.

Great care must be taken to maintain steady air pressure at all times. Pipejacking is often at shallow depths and in areas of disturbed ground. If cover is insufficient, air can be lost into fissures, old workings, into open granular layers or to the surface. This type of work should only be undertaken by skilled operators and after a full appraisal of the job. Working in compressed air can be a serious health hazard and most countries have regulations governing it. These must be strictly observed. Where operating pressures need to be greater than 1 bar, pressure balance shields offer a less hazardous alternative to compressed-air working. [See 3.2.4].

3.2.3 Auger-type shields

Auger-type shields are particular forms of mechanical shield which have to be operated remotely. Auger shields need to be distinguished from auger boring as such. The two operations have some similarities and auger shields draw their cutting and spoil removal technology from auger boring methods of installation. However, auger shields are independently-operated steerable units fitted onto the front of a range of product pipes.

Excavation is by a full-face rotary cutting head, connected and driven either by auger flights which remove the soil from the bore, or by a direct separate drive. The main types of head configuration can be divided into flight auger heads and the full-face type (drum, plate or dome). Flight auger heads are often spiral-shaped with a central pilot cutter or head, plus various arrangements of picks or cutters mounted on the head to cut the soil and drag it into the bore.

Auger shields can be further sub-divided into standard and pressurised types. They are all mainly used for smaller diameters, particularly for microtunnelling [See Chapter 5]. An example of a larger auger shield is seen in Figure 3.7.

3.2.4 Pressure chamber shields

Closed face remote-control machines, also referred to under the general heading of TBMs, incorporate a pressure chamber immediately behind the front cutting head or disc. The chamber, filled with water, slurry or soil, provides a counterbalance pressure to soil and water pressures. No access to the face is available and all operations are by remote control. In larger diameters the machine operator will be stationed at a control console in the rear of the shield. Figure 3.8 shows this arrangement. For smaller diameters the control panel can be located outside the tunnel. The overall layout of a pressure balance machine is shown in Figure 3.9.

Pressure balance machines have been designed to operate in unstable conditions such as non-cohesive soils below the water table. In practice, they are used in a much wider range of soil conditions. Many variations of pressure chamber shield have been developed, some of which incorporate crushing devices to cope with stones and boulders. Their types and variations are listed in Table 3.2.

Although pressure balance machines are made in larger diameters for traditional tunnelling, the upper limit for such shields in jacking is around 3500 mm.

Table 3.2 Pressure chamber shields

Type	Variation
Water	Standard Eccentric cone Crusher
Bentonite slurry	Standard Crusher
Earth	Standard Slurry Slime

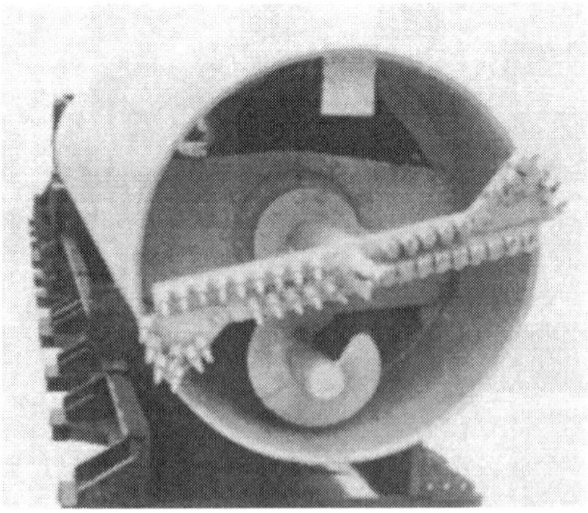

Figure 3.7 Bores head - auger shield with cutter bar (photograph by courtesy of American Augers Inc.)

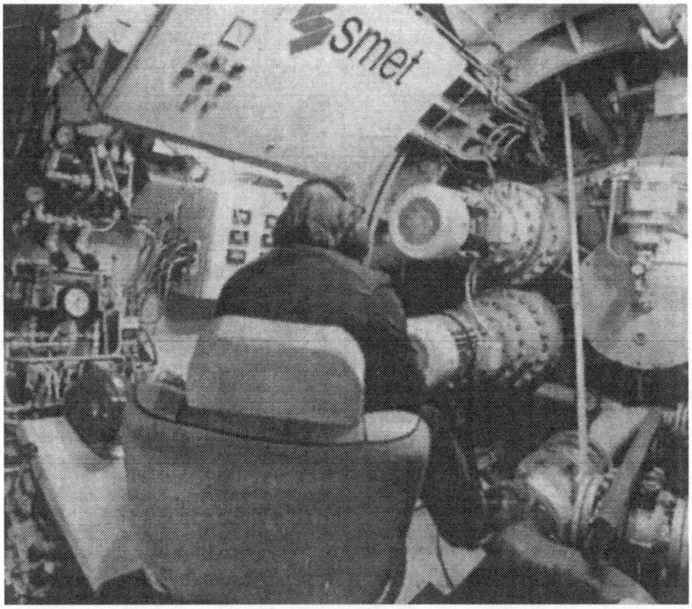

Figure 3.8 Operator in a large jacking shield (photograph by courtesy of Smet Boring)

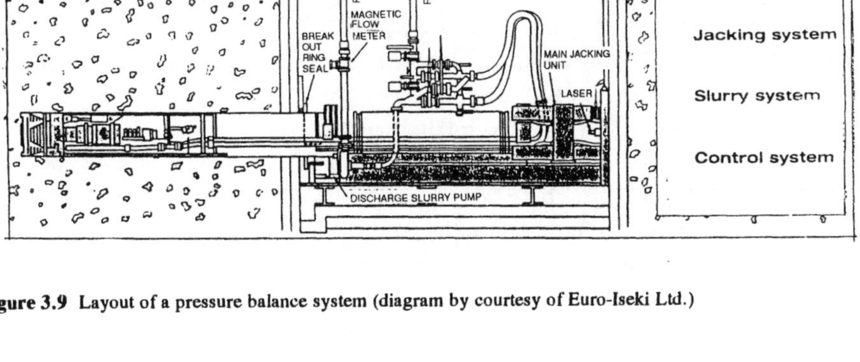

Figure 3.9 Layout of a pressure balance system (diagram by courtesy of Euro-Iseki Ltd.)

Water pressure-balance shields. Immediately behind the cutting head is a chamber where water is maintained at a pressure which balances the ground water level. These shields are designed for specific ground conditions. Frequently an auger-powered cutter head carries the excavated spoil back into the water pressure chamber, where it is mixed and transported to the surface by a pumped slurry system. For this type of water-based slurry, treatment and disposal are less complex and costly than for bentonite slurries.

Iseki's Unclemole, based on pressurised water balance, has been designed to work in ground containing quite large cobbles and heavy gravels. A cone crusher reduces all material to a defined maximum particle size before it enters the pressure chamber to be pumped away as slurry. The most appropriate soil conditions for this type of equipment appear to be cohesive soils with cobbles, and sands and gravels. Figure 3.10 shows the general arrangement of the Unclemole.

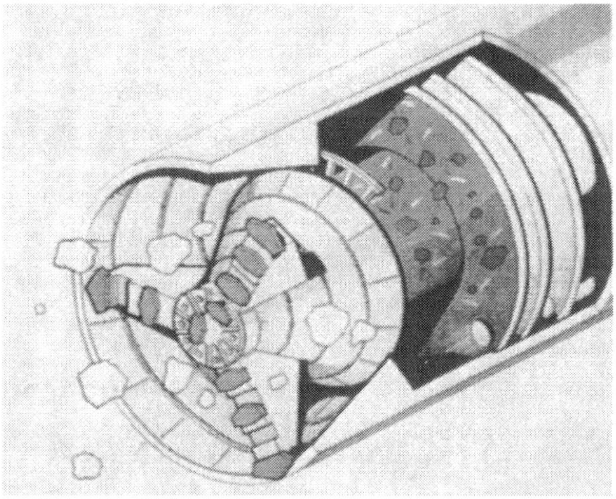

Figure 3.10 Iseki Unclemole (illustration by courtesy of Euro-Iseki Ltd.)

Bentonite slurry pressure balance shields. The original pressure chamber shields, introduced in the 1960s for traditional segmental tunnelling, were soon adapted for pipejacking. Their scale has been gradually reduced to the miniature versions used now in microtunnelling.

The face is excavated with a full-face rotating cutting head which has slits or ports through which the excavated material is forced into the pressurised slurry chamber behind the head. The openings are often controlled to regulate the amount of soil entering. The pressure inside the chamber is set to balance the ground water head. The slurry pressure is maintained by an incoming line from an injection pump as the outgoing line carries away the spoil cuttings suspended in the slurry. The slurry has a bentonite base but various additives may improve its stabilising performance. Cutting faces can be flat-faced plate, drum- or dome-shaped. Some versions of slurry TBMs also have a crushing capability. Stones and cobbles passing into the pressure chamber are diverted into a crusher chamber which reduces them to an acceptable size for pumping.

The difference between water-pressure and slurry-pressure machines and when either would be used are not readily obvious. For some cohesive soils it may only be necessary to use water, which, when mixed with the natural soil, forms a perfectly adequate slurry for face support and spoil conveyance. In non-cohesive fine-grained soils above the water table, the slurry will penetrate the face only a little, so no special measures are needed, except to ensure that excess slurry pressures do not cause soil fissuring. In coarse-grained non-cohesive soil, however, the objective is for the slurry to create a zone of penetration and to build up a filter cake. Depending on the nature of the soil, the slurry needs to be designed with appropriate additives, such as polymers and cellulose-based materials, if it is to fulfil these objectives. In open-structured ground, such as gravel or cobbles, even greater care needs to be given to the slurry formulation and its additives to avoid loss of material and pressure.

Earth pressure balance shields (EPBS). These shields use the excavated soil as the medium for containing the face soil pressure. Although mainly used in larger diameter tunnelling, some have been adapted for pipejacking. Some versions operate with a disc head and ports; others with spoke configurations. In the latter type, kneading paddles are mounted on the back of the spokes. How these are arranged and their *open ratio* are determined by the nature of the ground.

As the shield advances, the spoil is forced into the chamber behind, forming an impermeable plastic mass, which is maintained at a predetermined balancing pressure. The spoil is removed from this chamber by an auger arrangement. A variety of devices control pressure and spoil discharge to the required level.

The slime shield was developed by the Daiho company from Japan for working in high ground water pressure conditions or in sandy soils which do not flow plastically. A slurry based on bentonite or other chemical additives is injected into the chamber in relatively small percentages. This is then mixed in with the natural soil to produce the required plastic flowing material, which is also impervious. Figure 3.11 shows a Daiho *Slime* shield.

Combination pressure shields In several cases manufacturers have combined the advantages of pressure chamber and mechanical earth balance. The Iseki mechanical earth pressure counter-balance (MEPCB) slurry machine provides a degree of mechanical earth balance by having the cutting head mounted so that it can be independently pressed against the soil face. To do this it uses hydraulic rams to a predetermined force. It is claimed that, by using a combination of slurry pressure and mechanical earth pressure, the shield pressure can be maintained at a level above the active earth pressure but below the passive earth pressure. This reduces the risk of settlement and heave. It also allows a better control over the volume of soil removed by differential management of the position of the shield relative to the head and by adjusting the openings on the face disc. MEPCBs are particularly suited to working in water-bearing sands, silts and clays. Figure 3.12 shows the Iseki MEPCB.

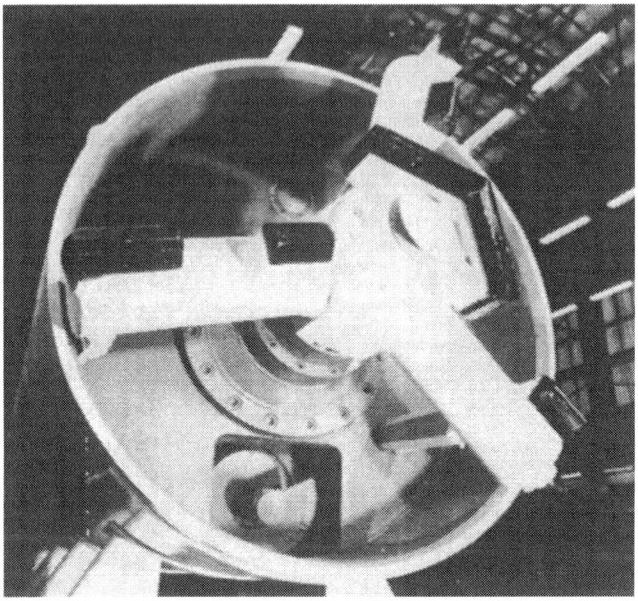

Figure 3.11 Daiho earth pressure balance shield (photograph by courtesy of The Dredging and Construction Co. Ltd.)

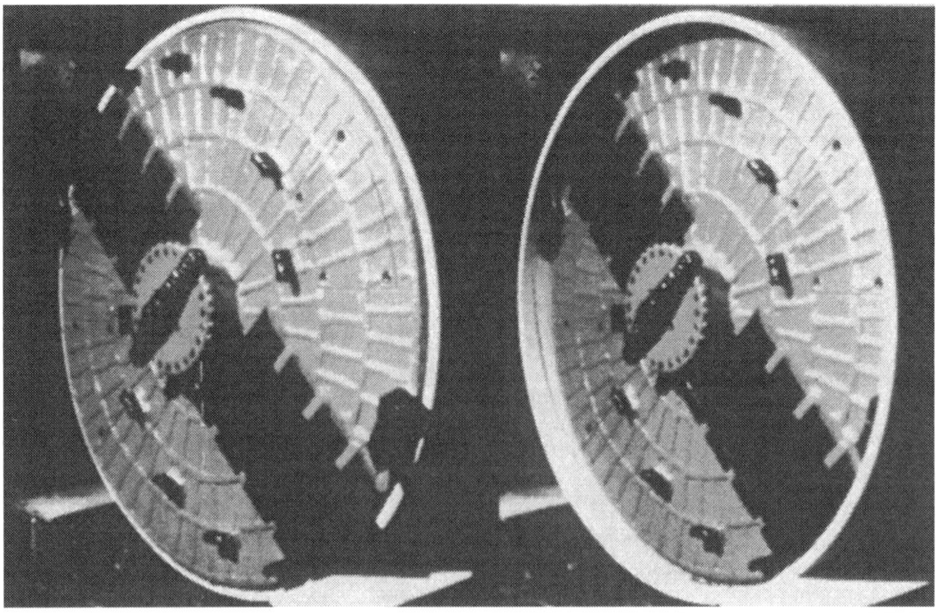

Figure 3.12 Iseki mechanical earth pressure counter-balance shield (drawing by courtesy of Euro-Iseki Ltd.)

In a similar way the Iseki Unclemole can be fitted with an earth pressure balance auger screw conveyor to take the spoil away from the chamber behind the crusher head for loading into carts or onto a conveyor. The layout of an earth pressure balance Unclemole is shown in Figure 3.13

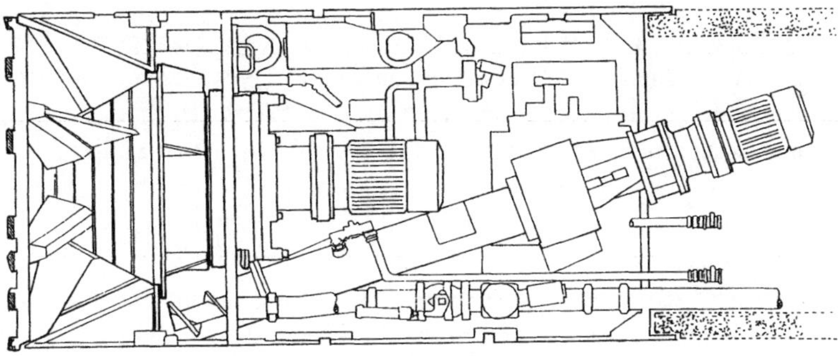

Figure 3.13 Earth pressure balance Unclemole (drawing by courtesy of Euro-Iseki Ltd.)

Air pressure balance shields. A fourth category of shield, not included in Table 3.2 because it has so far been used for larger tunnel shields, counterbalances the face pressures with air. The best-known example is the German Hydroshield, where a compressed air chamber is used as an *elastic spring element* with a slurry shield.

The concept of replacing slurry with air in a closed face remote-control machine has attractions and is being investigated in Japan. Komatsu already has a rotary-cutter remote shield working in compressed air. One big advantage is that the spoil is not polluted by slurry.

3.2.5 Design and construction of shields

Conventional open shields are frequently designed and constructed by specialist jacking contractors. Often the design is site specific. Mechanised and semi-mechanised shields are designed, manufactured and marketed by specialist tunnel equipment companies. Leading manufacturers in Europe include Herrenknecht and Westfalia Becorit (Germany) and Decon (UK) and in the USA, Akkerman.

Pressure balance machines are designed and marketed by a number of manufacturing companies in a range of sizes. Some specialise in one particular type of machine, others offer a range of types. Each company claims unique or distinguishing features for its products. Pressure balance shields have found their widest use in Japan, where they are offered by more than 20 manufacturing companies. Internationally, Iseki, based in Tokyo, but with sister companies in the UK and the USA, is a leader in the design and production of various types of pressure balance equipment. Other important Japanese companies include Komatsu, Okumura and Rasa. Herrenknecht is the leading European manufacturer. In North America, Lovat manufactures soft ground tunnelling equipment and has adapted its pressure balance shields for jacking.

3.3 Spoil cutting

The speed of the jacking operation is determined by the speed of excavation and spoil removal. Selecting the most appropriate method of face excavation is therefore important. Using the wrong face-cutting method can sometimes result in failure. The choice depends primarily on the anticipated ground conditions, but the length of drive is also a factor.

3.3.1 Manual excavation

The simplest, most flexible and, in many cases, most efficient method of excavation is by hand mining, as shown in Figure 3.1. For drives in stable soil conditions and over relatively short spans, experienced miners equipped with suitable air tools offer the most cost-effective route. Pipejacking is often used for crossing under railways and roads. Such crossings rarely exceed 100 m and are often considerably shorter. The time and cost of mobilising and setting up any form of mechanical equipment is prohibitive for such short drive lengths. A three or four man crew will be able to set up and start driving with an open shield in a few hours and can progress at rates of up to 10 m per shift for drives up to 1800 mm diameter in favourable ground. A typical 50 m crossing can be completed within a week. For smaller diameters, the productivity loss due to working in a cramped space is offset by the reduced volume of material to be excavated.

Man entry to the face is of particular value where natural or man-made obstructions are likely to be encountered. To be able to identify an obstruction and deal with it is a great advantage. Drives can also be made in a full face of rock by men operating air breakers.

Safety has always to be considered. In many countries man-entry working is limited to acceptable minimum sizes, most often 900 mm, and even then with restrictions on drive length. In Japan, as already mentioned, the smallest man-entry diameter is set at 800 mm. Limits to drive lengths in smaller diameters are determined by the facility of rescue for a face worker if he should be overcome or taken ill.

3.3.2 Partially-mechanised excavation

To speed up progress, manual excavation can be replaced by some form of mechanical device, such as hydraulic backhoes, cutter loaders and boom cutter arms. These were originally adapted from small trench back hoes and fitted with buckets or spade cutters suitable for working in the soft to firm soils. Although devices like these can be floor-mounted, it is often more convenient to suspend them from the roof of the shield, leaving the floor clear for spoil removal conveyors. Figure 3.14 shows a hydraulic back-hoe mounted in a shield.

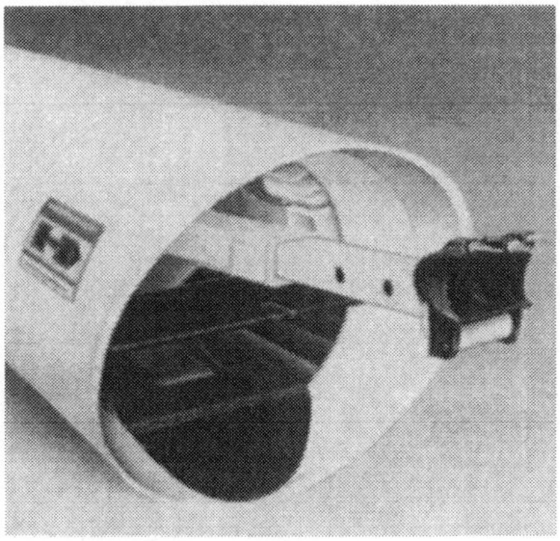

Figure 3.14 Shield with hydraulic back-acter (illustration by courtesy of Herrenknecht International Ltd.)

The cutter loader came from the coal mining industry and was adapted for use in pipe diameters mainly in the 1500-2500 mm range. It consists of a cutter drum which excavates the face and is driven by a scraper chain conveyor to transport the muck away. Harder material, including soft rock, can be cut in this way. Figure 3.15 shows this equipment.

Boom cutter arms mounted on a ceiling-suspended fulcrum are now widely used in shields for jacking and segmental tunnelling. The boom takes alternative forms depending on the soil conditions. A telescopic arm with a hydraulic-powered excavating bucket is suited for drives in soft to firm soils. For harder conditions, including rock, a road header boom can be attached. The rotating cutting head is equipped with an array of heavy-duty cutter picks which cut the face. Manufacturers have developed equipment with a common mounting, power supply and controls to allow the boom to be changed during the drive. Figure 3.16 is an example of a boom cutter.

Figure 3.15 Shield with cutter loader (photograph by courtesy of Westfalia Becorit)

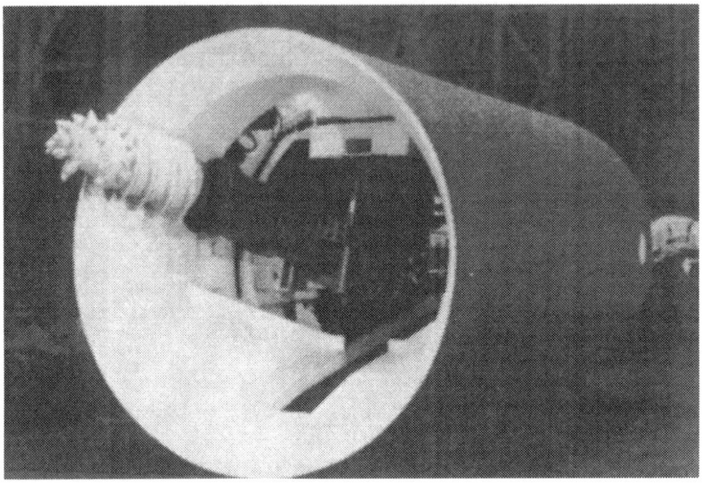

Figure 3.16 Shield with boom cutter (photograph by courtesy of Westfalia Becorit)

Equipment of this type combines the advantages of man entry to the face, flexibility of working in a wide range of soil conditions and relatively high production rates. Mechanical devices are available for operating in pipe diameters as small as 1200 mm and as large as 3500 mm. With these, installation rates vary according to ground conditions but can typically be 15 to 20 m per shift. Drive lengths can reach 500 m, or even longer for larger diameters.

3.3.3 Rotary cutting arms and wheels

A rotary cutting head is an alternative means of increasing excavation rates in relatively stable soil conditions. The simplest form, seen in Figure 3.17, has a single beam head powered through a central shaft. Cutting picks or chisels, mounted on the beam, sweep the entire face with a cutting action. The speed of rotation can be varied from 0 to 20 rpm, although most work is carried out in the lower range. Rotation can be in either direction to prevent or correct the tendency of the shield to roll.

Increasing the number of spokes from two to three, four, five or six will allow a greater array of cutters to be mounted. Cutters can be of the pick, chisel, button, roller or disc type. The increased cutting capability will need to be matched by greater torque and thrust. The central shaft bearing becomes a limiting factor and peripheral drive motors may be required. Drive motors on shields are either hydraulic or electric. In stable homogenous clays, production rates of more than 60 m per day have been achieved, although average rates are likely to be somewhat less.

Spoked cutting wheels are also used with pressure-balance machines. Figure 3.18 shows the head arrangement of an AVN 1500 from Herrenknecht.

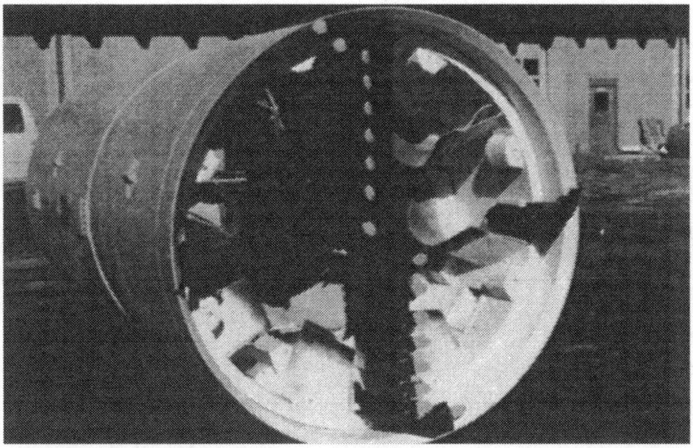

Figure 3.17 Shield with cutter beam (photograph by courtesy of Akkerman Manufacturing Inc.)

Figure 3.18 AVN 1500 with spoked cutting wheel (photograph by courtesy of Herrenknecht International Ltd.)

3.3.4 Full face machines

Full face machines can be of the open or pressure chamber type. The head can be a flat-faced plate, or drum- or dome-shaped. The head can be fixed in the shield or designed to move backward and forward in relation to the shield. The cutting face is usually equipped with a central pilot bit or head. Various bits, picks and drag teeth are arranged on the face, which can be used to cut soft to firm soil and to force it through the slits or ports located on the head. Hard-faced points, roller bits and disc cutters can be attached to break harder material like rocks and cobbles into fragments small enough to pass through the apertures.

Whereas worn or broken cutters of open-face machines can be changed, it is difficult if not impossible to change those of full-face machines. This can be a limiting factor on drives through abrasive soil conditions. A full-face cutting head for soft conditions and a head for use in cobbles and granular material are shown in Figures 3.19 and 3.20.

Figure 3.19 Full face shield for soft soils (photograph by courtesy of Euro-Iseki Ltd.)

Figure 3.20 Cutting head for cobbles and granular materials (illustration by courtesy of Okumara Heavy Industries)

3.3.5 Crushing heads

An alternative to breaking down cobbles, rocks and boulders at the face is to take them into the shield and crush them. There are two types of device: the cone crusher head or the crusher chamber. Both are designed to reduce material to small particles which can be carried away in the slurry discharge system.

The Unclemole incorporates an eccentric epicyclic cone crusher as part of the head. The eccentrically rotating cone head, which carries the cutting picks, forces soil including boulders up to 30% of the shield diameter into the shield face as it is driven forward. The material is radially crushed between the eccentrically rotating head and the cone-shaped wall, which is fitted with high-strength weld strips. Figure 3.21 shows the Unclemole crusher head.

The crusher-chamber types have scraper bits which force the material through ports in the face into a chamber, where, in the case of the Crunchingmole, a rotary crusher reduces the material to single-size particles. The Rasa crushing mole incorporates a jaw crusher. Figures 3.22 and 3.23 show these alternative arrangements. The size of obstruction that can be dealt with is limited by the aperture openings, but is about 20% of the shield diameter.

Figure 3.21 The Unclemole crushing head
(illustrations by courtesy of Euro-Iseki Ltd.)

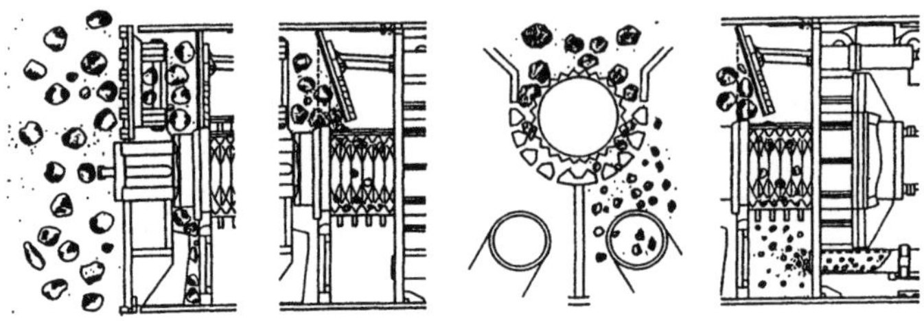

Figure 3.22 Mechanism of Crunchingmole

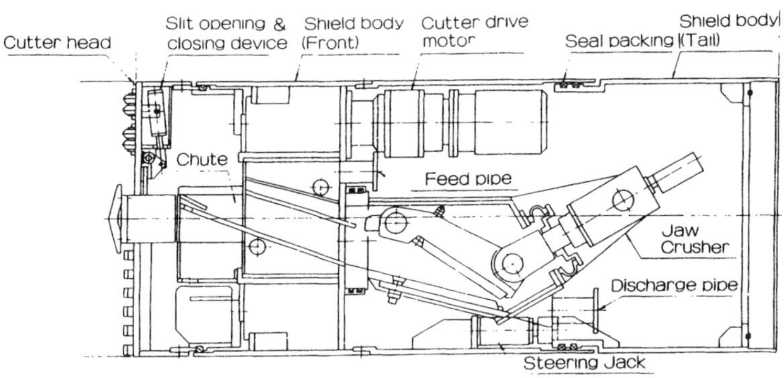

Figure 3.23 Crushing Mole — jaw crusher (illustration by courtesy of Rasa Industries Ltd.)

3.3.6 Rock heads

The growing demand for shields to work in much harder ground conditions has led to the fitting of rock-cutting heads to full face machines. Several manufacturers have developed special cutting heads for working in rock. Roller bits, carbide-faced teeth and button bits have been fitted on various heads. Disc cutters on larger tunnelling machines have proved very successful. However, scaling down cutters for smaller diameters presents some challenges to bearing design.

Rock heads are claimed to be able to work in rock rated as "very strong" (>100 mN per m^2). Figure 3.24 shows a rock head mounted with disc and roller cutters.

Figure 3.24 Rock head with disc cutter and roller cutters (photograph by courtesy of Euro-Iseki Ltd.)

3.4 Shield location, guidance and monitoring

3.4.1 Location

Basic requirements. A great deal of direct pipe installation by jacking requires that the pipe be installed to close line and level tolerances (±25 mm). Careful monitoring and close adjustment of the guidance system is required to achieve this accuracy. The scope and frequency of information that needs to be generated will depend on the nature of both the work and the equipment.

The fundamental survey information that has to be generated consists of the positional co-ordinates X, Y and Z

where	X is the designation for the horizontal co-ordinates
	Y is the designation for the vertical co-ordinates
	Z is the position along the longitudinal axis.

These co-ordinates are taken from a convenient frame of reference which is usually, but not always, the designated centre line for pipejacking and microtunnelling. In addition to the positional co-ordinates, the angles of rotation of the head are essential in some operations and useful in others:

Angles	*Roll* is the rotation about the main axis.
of	*Lead* is the rotation in the horizontal axis.
rotation	*Look-up* is the rotation in the vertical plane.

Roll is of special importance when the X and Y co-ordinates are not measured on the main axis. Lead (or *yaw)* gives a measure of how far one side of the shield leads the other. Look-up, also referred to as *pitch* or *overhang*, gives a measure by which the invert leads or lags behind the crown. Lead and look-up are not vital in locating the present position but are valuable in making predictions about the probable path.

The basic framework for any monitoring is the overall site survey; the established reference points and all monitoring should relate to that base.

Location and monitoring — man-entry. Surveying systems for man-entry pipejacking follow the same principles and much the same methods as are used in traditional tunnelling practice. The position and attitude of an advancing shield in relation to the required line and level needs to be monitored continuously, or at frequent intervals.

A surveyor with a theodolite is the classic way of obtaining the X and Y co-ordinates. Direct measurement or calculation from the number of pipes installed provide the Z co-ordinate. Roll can be measured directly in an open shield with the aid of a plumb bob, as can lead and look-up. Front and back line and level readings in the shield will also give the same information. Surveying by theodolite requires a skilled operator, is intermittent and interrupts working.

Laser monitoring, which overcomes many of these drawbacks, has become the preferred technique in pipejacking. A laser is set on a base or stand which has been previously surveyed and fixed. It will then provide a continuous beam of light and a bright spot on a target in the shield. This target is mounted in a manner that will avoid workers looking directly into the beam and where a clear unobstructed line can be maintained. A constant reference at all times is visible for taking measurements and for the jacking crew.

Lasers have some limitations in pipejacking, however. First, the laser mounting has to be towards the rear of the jacking shaft. Passive movement can occur in the surrounding soil when the rear jacking wall comes under load from thrust forces. The fixed position can therefore be moved, so giving inaccurate data. Self-grading lasers can compensate for tilting of the floor, but twisting and horizontal distortion need to be monitored by an above-ground theodolite or by a split-prism laser. By mounting a secondary target high on the front face, a secondary beam generated by the prism can be monitored for deviation.

A second problem is that deviation of laser beams can be caused by temperature differences along the line. These arise from heat generated by machinery, ventilation systems and even by direct sunlight on the pipe in the shaft.

A third problem is that over distance and in a dusty or misty atmosphere the beam becomes dispersed, which makes it increasingly difficult to get precise co-ordinate readings. Having double reference points through which the beam passes can help. For longer drives it is possible to mount intermediate reference points. By computing forward and backward readings the normal survey information can be provided.

For surveying curved jacking, a similar process is followed by locating a series of pairs of transmitters with direct lines of sight in the drive. Computer programmes can interpret and analyse data, and at the same time provide terminal readouts and a permanent record.

The laser-theodolite combines the two systems into one instrument, overcoming some of the disadvantages of the individual methods. Whilst operating with a laser beam, the theodolite provides a means of direct visual checking, including a check on any deviation or drift of the beam.

Gyro theodolites consist of a theodolite and a suspended gyro which identifies true astronomical North. Gyro theodolites have been used on tunnelling work and also on longer larger-diameter pipejacking [16]. Two things have inhibited their greater use: their cost and their susceptibility to vibration and shock.

The gyro compass assists in driving curved lines. One system uses a combined gyro and level set in the tunnelling shield. In addition, a stroke meter measures the shield jack extension, another meter measures pitch; and a third measures roll. A multi-task computer processes the information so gained, and calculates the position of the shield and its projected course. Shima [17] describes a gyro-compass incorporating a level device called a Gyvel. This has an accuracy of ±0.2 degrees in the horizontal direction and ±10 mm in the vertical direction.

Location and monitoring — non-man-entry. Where direct observation and measurement are not possible or where space is too restricted, the remote generation of survey information becomes a necessity. Devices like theodolites and lasers are still basic to the surveying function, but have to be supplemented with other measurement devices and also a means of relaying the information to a control console located remotely from the shield face. The mass of data has to be processed and interpreted by an operator, or computer, and the appropriate correction commands issued.

Remote-control monitoring is required for all remote-control jacking methods, including microtunnelling. As monitoring and guidance functions need to be closely integrated with the shield operation, manufacturers often supply the monitoring and guidance equipment as part of the shield package. This equipment also provides a print-out record of the drive and machine performance. Theodolite and laser techniques have been modified to meet the needs of remotely-controlled installation methods, particularly for microtunnelling Equipment for microtunnelling is discussed further in Chapter 5.

Closed-circuit TV (cctv) is widely used in remote-control microtunnelling and pipejacking. A control monitor receives camera pictures of the dials of instruments located in the shield, in particular the X and Y coordinates of a laser spot on the target. The operator's monitor and control board is shown in Figure 3.25. Active targets are of various types, but are based on laser-sensitive photocells which register the laser spot and provide a readout of the X and Y coordinates and the angle between the laser beam and the target axis. Figure 3.26 shows the layout for the target and cctv camera.

Most remote-control systems, such as cctvs and active targets, also incorporate inclinometers in the array. These can provide information on all three angular functions. Inclinometers are often incorporated into active targets, and with the use of sensors they can transmit information to the central control unit. Once processed and analysed, the data provide information on present position and attitude as well as a basis for predicting the path of the machine.

Some contractors still prefer the hands-on approach of having the operatives steer the head according to the information generated. However, the trend is towards fully automatic control based on computer analysis of data and computer provision of correction commands. Although some early automatic systems met some problems, recent experience has been favourable. Programmes need to introduce an element of *artificial intelligence* into computer decisions to avoid over-correction and to maintain a smooth path. Every machine has its own characteristics according to the geometry of the shield and the relationship of the cutter head to the shield face. Prevailing ground conditions also have considerable bearing on the response of a shield. Fully-automatic correction must take all these factors into account.

To artificial intelligence has been added *fuzzy logic* principles for processing and using stored data. A fuzzy logic system uses a database of known machine movements in a variety of conditions to predict the movement of the machine in the field, and then to provide the information needed for corrective action. At regular intervals it indicates the position of the tool head, deviations, grade and roll. A number of Japanese companies have led the development of fuzzy logic [18] [19]. One example is the Iseki Automatic Steering (AS) system, illustrated in Figure 3.27.

One of the drawbacks of laser target monitoring is that the target has to be situated some distance back from the face of the machine. All data therefore relates to this position rather than to its leading edge. Iseki has developed a modified system using mirrors and a split laser beam. One beam provides the required target line. The second beam is directed on to a reflector, which provides a second spot to give the predicted position of the centre of the drive head. By keeping the two spots superimposed, the operator can steer the correct course.

Figure 3.25 Monitor and control panel (photograph by courtesy of Laserbore Ltd)

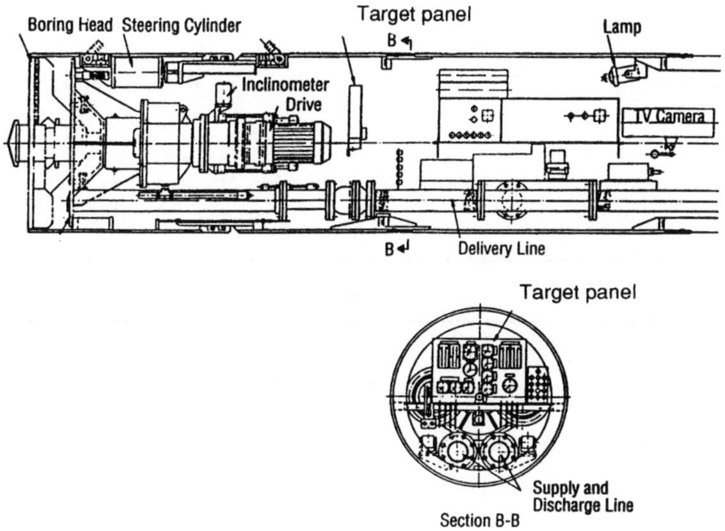

Figure 3.26 Laser beam, target and cctv camera monitoring (drawing by courtesy of Euro-Iseki Ltd.)

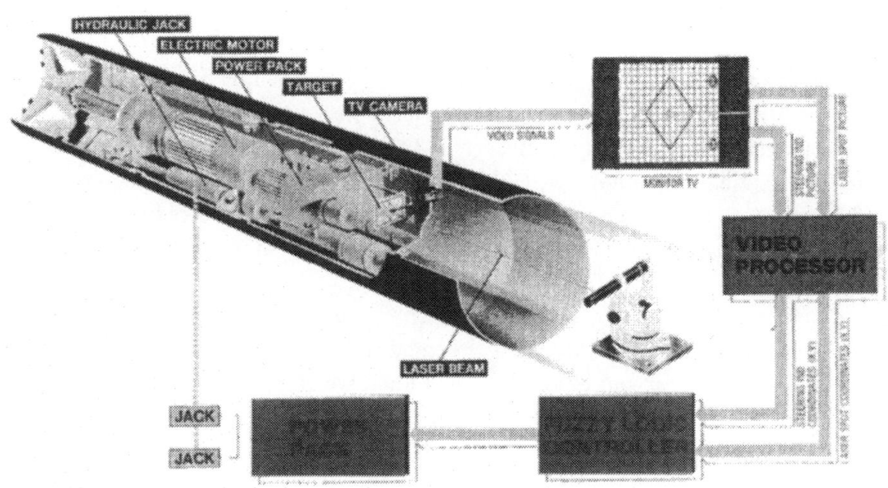

Figure 3.27 Automatic Steering (AS) system (illustration by courtesy of Euro-Iseki Ltd.)

Manufacturers claim that monitoring and control methods now available can detect deviations of 0.1 mm. These improved systems enable much tighter tolerances to be observed and generally allow contractors to work within ±25 mm with remote control systems over any length of drive. In some cases tolerances of ±5 mm are quoted.

Other survey methods. Hydrostatic gauges can monitor level (Y coordinates) and pipe grade. They detect differential liquid levels and relate them to known reference points. A sophisticated version measures the change in hydrostatic pressure by using a diaphragm on a sensor which transmits information electronically along the connecting tube back to a digital display monitor.

Detection systems based on electromagnetic induction methods have found limited use in pipejacking. The horizontal position of the shield is calculated from the strength of the magnetic field generated by an electronic coil mounted in the shield. The signal is measured by a receiving coil located at the surface. Vertical position can be measured by the pressure sensors described above. One shortcoming is that electromagnetic induction systems have a limited operating depth of around 6 m. Another is that the electromagnetic signal can be dispersed by adjacent buried objects. Achievable accuracy of monitoring is often insufficient to meet tight tolerances.

An optical fibre gyroscope developed by Nippon Telegraph and Telephone (NTT), was described in a paper given to the Paris No-Dig Conference in 1992 [20]. This was designed to detect small changes in the angle and direction of driving. The optical fibre gyroscope appears to overcome some of the drawbacks associated with the traditional gyro-compass. A monitoring system of this type would offer advantages, particularly in curved driving.

3.4.2 Guidance

The steering characteristics of conventional shields are controlled by the shield geometry, which in turn is influenced by soil conditions. A shield steers more easily when its ratio of length to diameter is around unity. However, in soft soils, to obtain the deflecting forces necessary for steering corrections to be made, a greater length of shield is required. The ratio of length to diameter may increase to 2, or even 3. This ratio is reduced by fitting double-action steering jacks into the shield. All but the most simple shields now have this type of jack.

Simple shields will be constructed in one section with angular adjustment occurring between the shield and first pipe. More sophisticated shields are constructed with lead and tail sections connected by means of an articulated joint, which is fitted with a seal to allow differential angular movement of the lead section.

Cylinders have a stroke of between 300 and 600 mm where the rams are dual-function, for independent shield thrusting and angular adjustment. Independent thrust capability has several uses. It can reduce the total jacking load by separate forward movement of the shield. It can help stabilise the face in soft and unstable soils. In rock and tough soils it can provide thrust to the cutters. Where the rams only perform the adjustment function, cylinders with a much shorter stroke and smaller thrust capacity can be used.

The positioning, number of rams and the jacking capacity vary according to the diameter of the shield and the manufacturer's design. A common layout uses four rams mounted at the 45° points; other arrangements use three. Larger diameter shields may require six or eight control cylinders. In conventional open shields the differential setting of the shield rams can be done directly by operatives at the face. In remote-control shields solenoid valves are used to control the setting remotely.

The hydraulic ram steering system can be used to correct pitch and lead deviations but has no effect on shield roll. The use of clockwise and counter-clockwise rotation during working can help arrest the tendency for the shield to roll. Some shields are also fitted with hydraulically-controlled stabilising fins set in the sides of the shield.

3.4.3 Monitoring additional information

Apart from the results of the different survey coordinates, some additional information needs to be monitored and recorded in remote-control operations. This includes:

- Cutting face stability
- Machine performance
- Spoil removal
- Changes in ground conditions

Face stability can be monitored through pressure cells in the head and pressure chamber and also by monitoring changes in soil weight. Machine performance, such as torque, hydraulic pressures and shield ram position, can be monitored along with temperature. In slurry transportation systems, operating pressures and volumes can be recorded.

A valuable practical device incorporated in a number of manufacturers' packages is an acoustic device, which records the sound of the cutting head. From the noise, an experienced operator can quickly recognise any change in ground conditions or when an obstacle is encountered. The operator will be able to detect the difference between rock or cobbles, timber or metal.

4 Pipejacking equipment: line, drive pit, top side

4.1 Spoil removal

Spoil arising from excavation of the face has to be transported away along the line to the drive pit, and from there to the surface where it is removed from site. Removing this spoil is often the most critical part of the operational cycle and can be the key to the rate of progress in pipejacking. This is not so for microtunnelling. One main limit on increasing the capacity of spoil removal is the small cross-sectional area of the pipe, which allows no opportunity for a by-pass with wheeled systems, except in the largest pipe sizes. For this reason pumped systems of spoil discharge are frequently used, even in situations where ground conditions do not demand a pressure balance chamber.

The system for transporting the spoil has to be compatible with the nature and volume of excavation coming from the shield and with the drive length. Depending on these two factors, there are several ways to dispose of spoil from the tunnel:

- In skips on wheels or rails
- By conveyors
- By positive displacement pumping
- By slurry pumping
- By screw conveyors
- By vacuum extraction

The first three of these methods are used in man-entry pipejacking. Slurry pumping is used extensively both in pipejacking and microtunnelling, while screw conveyors and vacuum methods have their greatest use in microtunnelling. Whichever method is selected, it is important to ensure that the spoil disposal system is integrated with the rest of the equipment and the operation as a whole.

4.1.1 Wheeled systems

For short man-entry drives, manual handling of wheeled skip systems is possible, but it is labour-intensive and tiring. It is inefficient on long spans and unsafe on significant inclines.

Winch-powered double drum *endless rope* haulage arrangements for transporting skips in and out of the pipe are an improvement on manual handling, but tensioned cables can be a hazard to workers if safe working procedures are not strictly observed. Winching speeds of around 5 to 10 km an hour are typical, although higher speeds are achieved. Winching is reasonably efficient over drive lengths of up to 250 m.

With increasing drive lengths, rail tracks and battery-powered locos can be cost-effective, but unless purpose-designed they can be wasteful on space. If rail systems are used, the skip wheels can be mounted within the body of the skip to make the best use of vertical space. Figure 4.1 shows a loco skip used in pipejacking. Equipment of this type can work in pipe diameters as small as 900 mm and achieve speeds of up to 4 m per second. A conveyor is often installed at the face for loading the skips.

Figure 4.1 Scharf loco skip (photograph by courtesy of Maschinenfabrik Scharf)

Further space savings in pipes can be made by using cars with angled wheels to run on the pipe itself. Wheels are of rubber to prevent damage to the pipe. This arrangement allows the greatest use to be made of the pipe profile in maximising spoil removal.

At the shaft, cranes lift the detachable skips to the surface. Removing the full skip for emptying, and returning the empty ones holds up the operation. To alleviate this, bunkerage can be provided by a short siding at the opposite side of the shaft or by using a secondary cranage device to drop in an empty skip as soon as the full one is lifted.

4.1.2 Belt and chain conveyors

Short belt and scraper chain conveyors can be effective for loading skips or feeding hoppers for other means of spoil transport. However, it is difficult to employ these over the total drive span as the main conveyor structure, necessary for stability and direction, will take up a large proportion of the cross-sectional area. Some form of spillage is usually inevitable, which will slow down productivity and threaten safety, particularly in confined spaces.

When large cross sections are being jacked over short spans, the large carrying capacity of conveyors has proved effective and economic. At the shield, narrow-profile extendible overhead conveyors are popular for loading skips as they can reach several skips in one cycle. On longer drive spans the saving in time becomes more significant.

4.1.3 Positive displacement pumps

The positive displacement pump for muck removal, originally designed for pumping concrete, was first used in Japan in the 1980s for traditional large diameter jacking. Since then it has been scaled down to a size which can be operated from a pipe diameter as small as 1400 mm.

Soil from the shield is fed into a mixing trough and mixed to a suitable consistency. A small amount of water, bentonite slurry or polymer can be added to increase pumping efficiency. A screw conveyor carries the mixed material into the closed pump hopper. A double acting piston pump then forces the spoil along the discharge line at a pressure up to 120 bar.

The spoil needs to have a consistency similar to wet concrete and must contain a minimum of 400 kg/m^3 of material less than 25 micron. Where soil is not well graded and contains rock pieces this would have to increase to about 700 kg/m^3. The diameter of the delivery line should be three times the size of the normal maximum aggregate size. Single particles can be up to 90% of the line's diameter.

A Putzmeister TMP 16 will work inside a 1500 mm pipe and can pump up to 16 m^3/hr. This is equivalent to an excavation rate of 105 mm per minute, or installing a 3-m length of 1500 mm pipe in 30 minutes. Figure 4.2 shows a Putzmeister pump installed in a 1500 mm pipe.

Positive displacement pumps are capable of working continuously over long lengths and pumping the spoil up the shaft and directly into hoppers. Another advantage is that the spoil does not require elaborate separation processes before disposal.

4.1.4 Pumped slurry

Pumped-slurry spoil removal is preferred with many microtunnelling systems and pipejacking shields. One reason is that the slurry circuit usually provides the means for ground control at the face. Such systems are also economical on cross-sectional area occupied, as they can carry high volumes of material over long distances through the pit and to the surface.

A pumped system consists of a charging line which carries mud to the suspension chamber located behind the cutting head, and a discharging line which carries the slurry away. *Mud* is defined as the fluid which is designed and mixed according to the conditions. *Slurry* is defined as the mud containing the excavated material. In practice this distinction is rarely made. Either water or bentonite mud can be used as the conveying medium. Water has the advantages of being cheap, readily available and easier to separate out from soil. Clay and silts work well with water, but a bentonite or polymer mud will be required with sands and gravels to provide the support function at the face. The density of the slurry that can be pumped is limited to about 1.4 tonnes per m^3. The maximum load that can be carried in suspension is therefore about 400 kg/m^3. Provided the shield is equipped with some sort of crushing mechanism of adequate capacity, a pumped slurry system can cope with most types of ground.

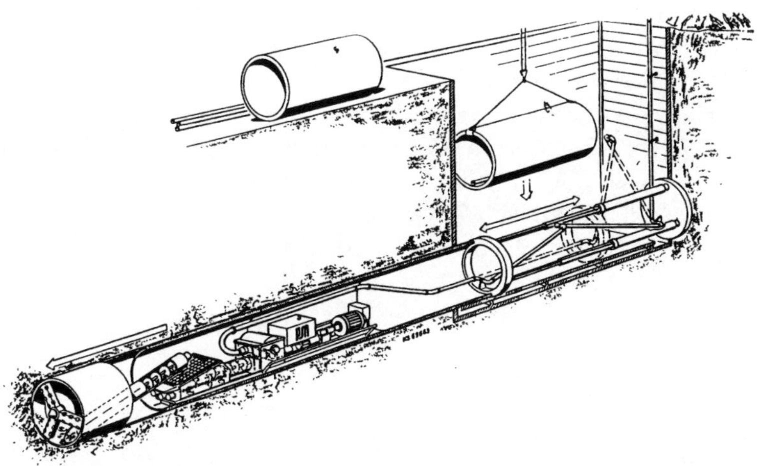

Figure 4.2 Putzmeister TMP 16 pumping system (drawing by courtesy of Putzmeister Werk)

The correct match of pipe diameter, slurry concentration and hydraulic velocity are essential to the effective design of systems. The working velocity needs to be greater than the critical velocity. The critical velocity of heavier bentonite-based slurries is less than for water. Flow and density measuring equipment is available although many contractors use standard viscosity checks plus the experience of the operator. A much fuller review of the theory and design of hydraulic pumped systems is to be found in Dr. Stein's book [21].

At the drive shaft, provision has to be made for accommodating the jacking sequence, which involves adding successive pipes and pushing them through the pit. Shut-off valves keep the line under pressure, while quick-action couplings facilitate breaking and remaking as each new section is added. Charge and discharge lines are flexible connections at the shaft bottom to allow for the movement of the pipe as it is jacked through the pit.

4.1.5 Screw conveyors

Screw conveyors (also known as augers) working within a steel casing pipe are a favoured method of spoil transport in microtunnelling and are covered in Chapter 5. Their use in pipejacking has so far been limited. Moving larger volumes of variable material over longer lengths creates problems with power requirements, and often causes shaft and coupling failure as well as blockages.

An example of a screw conveyor used in pipejacking is the Westfalia WBM-L12, illustrated in Figure 4.3 The equipment is designed for installing pipes up to DN 1200 mm. The 120 mm screw conveyor running in a casing is small enough to allow man entry into the shield. The cutting head is powered by a separate drive. Screw conveyors will move the soil to the shaft bottom where it normally discharges directly into a skip. The removal and discharge of skips suffer from the problems discussed in 4.1.1.

When using any form of screw conveyor it is important to regulate the volume of material entering the flight augers to prevent their overfilling thereby causing considerable increase in friction load and the risk of seizure. One way in which the volume can be regulated is to fit a tighter pitched lead auger.

One very specific use of screw conveyors is for earth pressure balance shields, where short sections move spoil from the pressure chamber into a spoil removal system. To maintain the earth pressure in the chamber, the screw conveyor speed is carefully controlled to balance soil removal with forward progress. The conveyor is also fitted with a gate valve or similar device at the outlet port to cut off water flow.

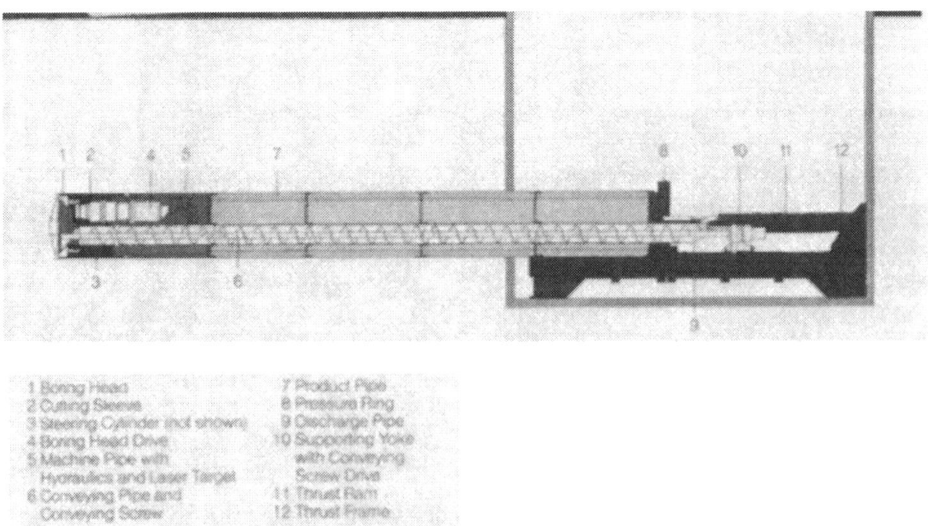

Figure 4.3 WBM-L12 auger mucking system (drawing by courtesy of Westfalia Becorit)

4.1.6 Vacuum extraction

Pneumatic conveyors, both pressure and suction types, are employed in a number of industries. Some microtunnelling manufacturers have adopted vacuum suction systems. These are described in Chapter 5.

The perception is that the amount of power needed to move the volume of spoil from a pipejacking operation makes vacuum extraction impracticable and uneconomic for most situations. However, a German vacuum system for spoil removal uses suction pipes in sizes from 50 mm to 250 mm and also a special pump which gives up to 80% vacuum. Material is loaded into a hopper at the face, sucked along the line and discharged into a closed tank. Extraction distances of over 200 m are claimed, with outputs of 100 tonnes of material per hour. An air supply of 11,500 m^3 per hour is required. However, the types of material that can be extracted are limited, as cohesive material can clog the line. Excess water can also be a problem.

4.2 Jacking equipment

The load necessary to move the line forward is provided by a series of rams or jacks pushing successive sections of pipe through the drive shaft. On longer drives the thrust can be supplemented and distributed by intermediate jacking stations (IJSs) located at strategic points in the line of pipes. The design of shafts and the reaction walls is considered in detail in Chapter 9.

4.2.1 Jacking rigs

The main elements of the jacking rig are the jacks or rams, the framework and the thrust ring. The jacks are almost invariably hydraulic cylinders. The framework locates and supports the jacks and transmits the jacking forces to the reaction wall and also onto the thrust ring.

The original rigs were often put together using material and components readily available in a contractor's depot and were often crude and inefficient. Current rigs are designed to be integrated with the overall operation and to occupy minimum space to keep down working shaft dimensions. The speed of jacking needs to be matched to the speed of excavation and progress. The rig layout has to facilitate pipe and skip handling as well as monitoring and control devices. Rigs have also to be easily set up and adaptable for a range of work. Some designs have the power unit, providing hydraulic and electric power to the various components, incorporated in the jacking rig. A more common arrangement is to situate the power units away from the rig. A typical arrangement for pipejacking is shown in Figure 4.4. Jacks can be of any number from two upwards. The total amount of jacking capability provided is related to the safe load that can be applied to the section being pushed, plus some margin to accommodate high start-up loads. For driving large box sections the number of rams may be more than 20 and have a total pushing capacity of several thousand tonnes. Individual hydraulic cylinders are normally operated at high pressure between 300 and 650 bar and can have a jacking capability from 30 to 300 tonnes. Cylinders are always power-retracted and some are fully double-acting. Power packs providing the hydraulic power are sized according to the number and size of cylinders and the speed of jacking.

As the pipe is pushed through the pit the jacking load has to follow. This can be achieved in a number of ways. Two- and three-stage telescopic cylinders are available which will provide strokes up to 4000 mm, which is more than sufficient to jack most pipe sections. These cylinders, normally fixed at the rear of the shaft, have a substantial basic length when retracted which has to be accommodated by an increased length of pit. See Figure 4.5.

A second approach is to arrange that the frame, carrying short-stroke jacks of, say, 300 mm, travels along a sub-frame. The jacking frame is located on and locked to the sub-frame by a suitable pocket-and-pin device. After each stroke is completed, the frame is unlocked, moved forward and relocked. This type of rig can be seen in Figure 4.6.

A third approach is to maintain the cylinders in the same position and use mechanical spacers after each stroke to extend the thrust. With multiple spacers, some form of locking is required to avoid the danger of jack-knifing. Figure 4.7 shows a rig with spacers formed from half-round sections.

Many jacking rigs incorporate a steel back plate to ensure even distribution of the jacking load from the rig onto the reaction wall. The thrust collar, or ring, which spreads the load uniformly on the end of the pipe needs to be rigid and to fit closely to the end profile of the pipe. Packing material on the pipe face helps to distribute the load evenly. This is either cast or fabricated as a box section in steel. Figure 4.7 shows a typical thrust collar. Thrust cylinders are fitted with ball-and-socket end plates to ensure uniform transmission of the jacking load on to the collar.

Figure 4.4 Telescopic cylinder jacking rig (photograph by courtesy of Akkerman manufacturing)

Figure 4.5 Akkerman frame and pin jacking rig (photograph by courtesy of Akkerman manufacturing)

Figure 4.6 Jacking rig with half round invert spacers (photograph by courtesy of Thrust Bore Ltd)

Figure 4.7 Thrust ring (source, Akkerman Manufacturing)

4.2.2 Intermediate jacking stations (IJSs)

IJSs are used to supplement the jacking load provided at the shaft. As already noted, it is the safe load on the pipe which determines the jacking capacity in the shaft. For longer drives, or in certain types of difficult ground, the jacking loads will increase beyond the safe limit. It is therefore necessary to spread the load over a number of points to reduce the jacking load at any one point. This is done by inserting IJSs in the line. A typical layout of an IJS is shown in Figure 4.8 and the arrangement of cylinders in Figure 4.9. With the pipes to the rear forming the reaction block, the front part of the line is jacked forward over a distance of 300 to 600 mm. The jacking capacity at a station is determined by the safe load that can be applied to the pipe.

The operation of an IJS is illustrated in Figure 4.10. An outer steel casing or sleeve is fitted to the trailing end of a pipe. A special recessed pipe is required at the leading edge on the first of the following pipes. As the station is operated, the sleeve slides up and down on the recess. A rubber sealing ring between sleeve and recess stops ingress of water and fine material. The load is transferred onto the leading section by a fixed ring and the reaction on the pipes behind by means of a loose ring. When the drive is completed the hydraulic equipment can be removed and the station closed up. Rams used in IJSs need only be single-acting, as they are closed up by the thrust applied down the line. To ensure the load is spread uniformly, cylinders with ball-and-socket spreader plates are recommended.

Operating a number of stations can be programmed to take place in a series of automatic sequential steps without hindering the normal operation. As distance increases it will no longer be feasible to operate the IJSs from the main power pack. Small hydraulic power packs which can serve several stations are therefore fitted in the line. These *slave* stations will still be controlled remotely from the central control panel.

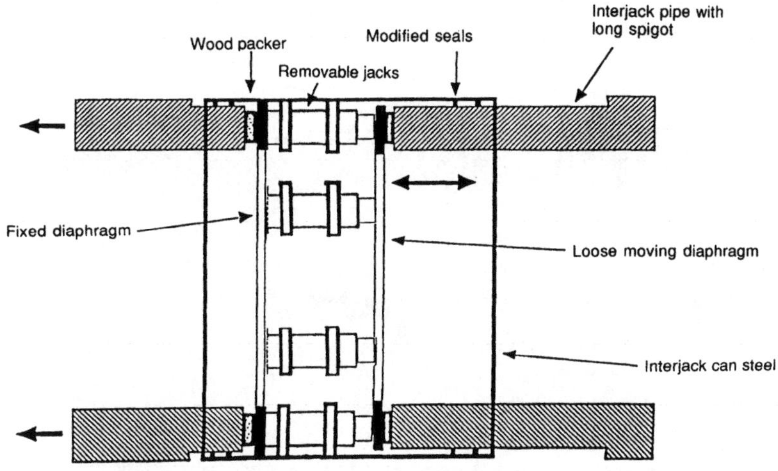

Figure 4.8 Layout of an IJS

Figure 4.9 Hydraulic rams at IJS (photograph by courtesy of Westfalia Becorit

Increasingly it is becoming good practice to install an IJS in the line at a relatively short distance behind the face as an insurance against high loads being unexpectedly encountered at the shield. Should this happen, this station can be put in operation.

Both the location and the number of stations in any drive are again determined by the estimated jacking forces and by the safe loads on the pipes being driven. In long drives a considerable number of stations may be needed. For the Europipe Landfall Tunnel in North Germany a distance of more than 2500 metres has been installed in a single drive with the aid of 25 stations.

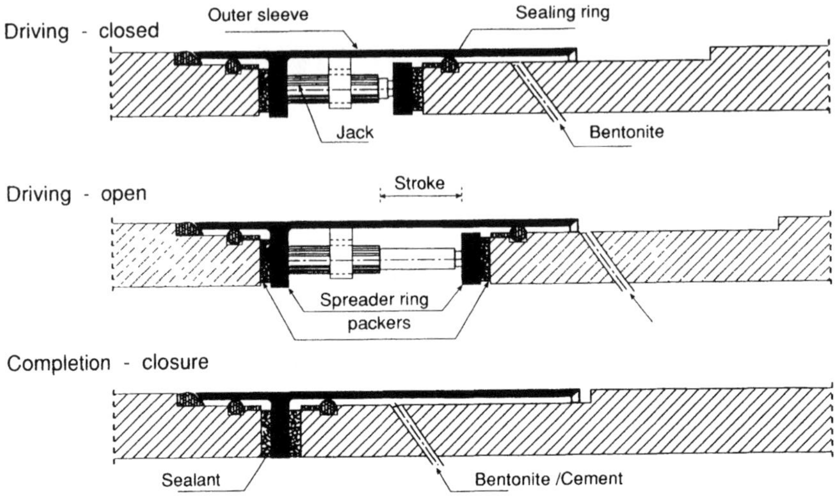

Figure 4.10 Operation of an IJS

The positioning of IJSs should be planned before jacking commences and all necessary stations and special pipes procured. It is advisable to be generous in determining how many stations will be needed. The cost of installing an additional IJS at the start is relatively small, even if it is not needed because jacking forces remain low. However, if stations have not been installed and loads suddenly increase, they cannot easily be retrofitted into the line. On one occasion when an IJS was fitted in the middle of a steel line, the line had to be cut on site and an IJS had to be fabricated in-situ. In every case, jacking loads should be monitored carefully. If they are rising above predicted levels, it may be necessary to install stations sooner and at closer intervals.

The concept of IJSs is extended to its ultimate in the Unitunnel system, developed in the late 1970s by M. Richardson and illustrated in Figure 4.11. The Unitunnel uses a "station" between every pipe and an inflatable bladder instead of hydraulic rams. The bladders are automatically inflated and deflated in sequence. The object is to reduce jacking loads to negligible amounts and in theory to drive over long distances. Variations on the theme of using a large number of automatically sequenced IJSs have been proposed by several people [22].

The Unitunnel method won the Civil Engineering Innovation Competition in 1981 sponsored by the magazine New Civil Engineer and the National Research Council. The judges thought it "a brilliant idea" and a significant development that filled an "urgent need" in the industry [23]. The rights were sold to a leading British tunnelling contractor, but after a number of problematic jobs they lost interest. Today the Unitunnel Association in Japan occasionally uses a modified version for which the main application appears to be driving curved lines with a radius as small as 31 m.

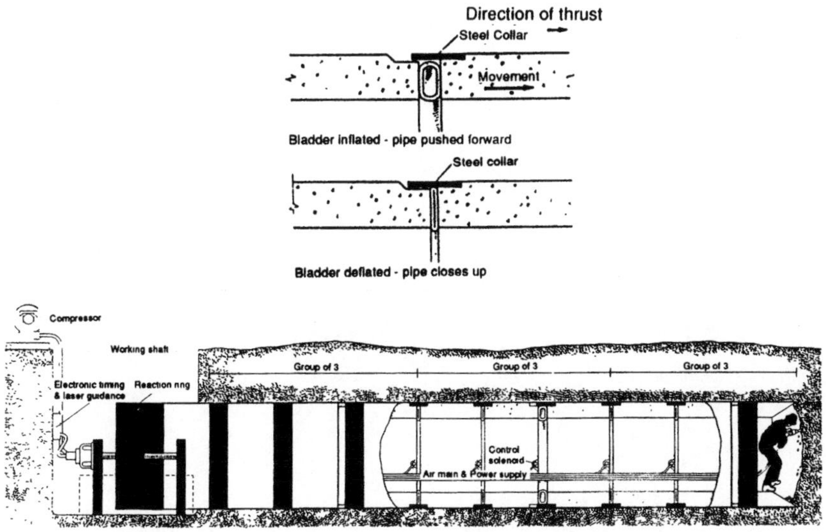

Figure 4.11 Unitunnel system (after West, G. Innovation and the rise of the tunnelling industry)

4.3 Top side - surface equipment

A number of support functions and services have to be provided in the area around the head of the drive shaft. These include:

- Spoil handling and disposal
- Pipe handling
- Pipe storage
- Primary power units - electric, hydraulic and pneumatic
- Mud mixing and charging equipment
- Lubrication mixing and injection equipment
- Control stations

4.3.1 Spoil handling and disposal

Where skips are being used in conjunction with wheeled or auger methods of disposal, the muck is hoisted by the site cranage and tipped into overhead hoppers and then into waiting wagons or containers.

With a pumped slurry system, separation facilities are needed to remove the spoil from the water or mud. The volume of spoil generated by slurry machines can be as much as ten times the net volume of the bore excavation. As the problems of its treatment and disposal, together with their associated costs, form a significant part of a jacking operation, slurry systems warrant close attention. Figure 4.12 shows a typical site arrangement.

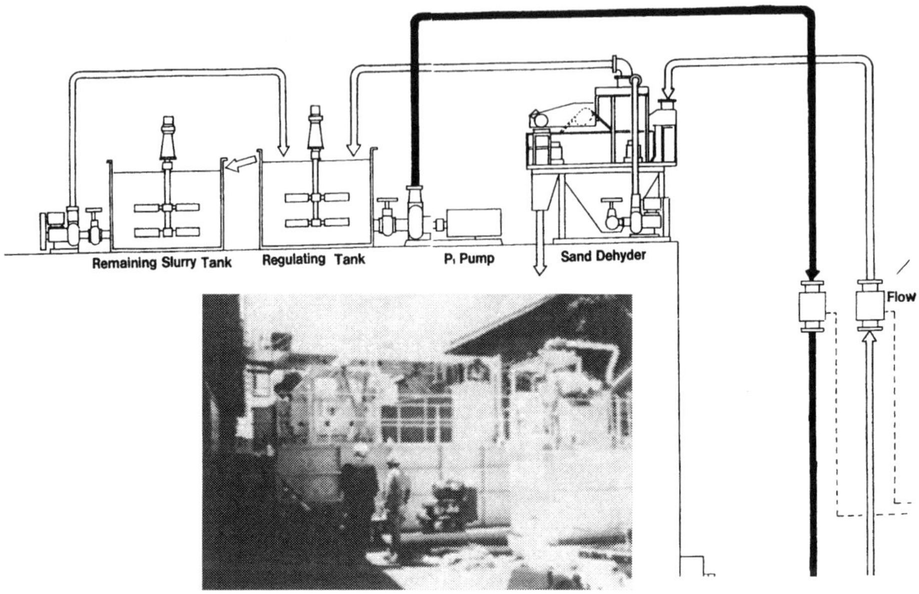

Figure 4.12 Rasa site separation plant (source Rasa Industries Ltd.)

The equipment for mud mixing and charging are normally closely integrated with the slurry separation plant. The aim is to separate the mud mix from the solids and to re-use it as the carrier fluid. Separation methods are various:

- Settlement tanks
- Screens for coarse material
- Hydrocyclones and centrifuges
- Dewatering screens
- Filter presses

These processes can be used singly and in combination, depending on the material that has to be separated, the space available, and also the regulations on disposal of contaminated soils. The size of the process plant is matched to the diameter and speed of driving. For pipejacking in diameters from 1000 mm to 3000 mm, the capacities required are likely to be in the range 10 to 200 tonne/hr.

Settling tanks offer a simple and cheap solution where the carrier fluid is water. A typical contractor's set-up will consist of a prefabricated steel tank, or several in a connected series where longer settlement times are needed. Settlement tanks are effective for separating gravels and sands, but only particles above about 1 mm sink in water (without help) quickly enough to make it practical to use simple settlement tanks only. The addition of flocculants can speed up the sedimentation process where the density of the slurry is not high. These flocculants are synthetic, organic, water-soluble polymers which need to be added to the water at a ratio of about 0.1% of the volume to be treated. Settling tanks may not be possibilities on sites where space is limited.

Screens to separate coarse materials are often used as the first stage of a multi-stage plant. A surprisingly large amount of material can be screened off. For instance, many clays are cut and transported in lumps. A range of screen types is available in a variety of materials with different surface areas and screen openings. The appropriate choice is likely to need the advice of a specialist.

Hydrocyclones, which are based on centrifugal separation principles, can separate material down to 100 microns in a single-stage plant and down to 35 microns in a two-stage plant. These are cheaper to purchase and operate than centrifuges. Centrifuges with a helical conveyor can extract fine particles down to 5 microns and smaller with the addition of flocculants. The density of the incoming slurry has a considerable bearing on the performance. The outputs from hydrocyclones and centrifuges have a high water content and have to be transported in containers, not welcomed at many tipping sites.

Vibratory dewatering screens are used for reducing water content. Dewatering belts and filter presses reduce the water content of the finer soils by about 20% to 25%, allowing it to be transported in conventional trucks. Dewatering by belts and presses is expensive and is avoided where possible.

Several leading manufacturers offer proprietary separation plants of modular design, which can be adapted to ground conditions. Some specialist manufacturers of separation systems also serve large-diameter tunnelling and other civil engineering and geotechnical uses of bentonite slurries. Separation plants represent a high proportion of the capital investment in a microtunnelling or slurry shield pipejacking system and should be selected with care. Meseck [24] and Anheuser [25] give more detailed information on separation and treatment plants. If it proves uneconomical to try to remove finer particles, it may be better to remove the slurry from the site by tanker for disposal at a controlled tip. But finding a tip can sometimes be difficult and charges are high. The loss of the bentonite is a further expense. The rising costs of slurry treatment and the problems of disposal are weighing against slurry-pumped systems in some countries.

4.3.2 Pipe and other material handling

The initial positioning of the shield and jacking rig and the subsequent placing of pipe sections require some form of cranage. Mobile cranes are widely used and normal care needs to be taken that the heaviest item to be handled is well within the safe lifting capacity. On many sites expensive mobile cranes can prove uneconomic, particularly if they are under-utilised. A gantry alternative can be economic for longer drives and larger pipe diameters. Figure 4.13 shows a gantry crane operation. For smaller diameter pipejacking and microtunnelling equipment, a gantry crane now often forms part of the equipment package.

Restrictions on space often limit the number of pipes that can be stored on site. A secondary storage area nearby may be required to allow frequent site re-stocking. Pipes are prepared with joint packers fixed to the joint face before they are dropped into the pit. Any temporary pipes and cables, such as the discharge and charge pipes for slurry pumping, are also positioned. Pipe slinging arrangements must be attended to. Over and above meeting relevant safety requirements, they need to be designed to facilitate rapid and accurate placing at the pit bottom.

Figure 4.13 Gantry crane (photograph by courtesy of Kennedy Construction Ltd.)

4.3.3 Power supplies: air, hydraulic and electric

Various items of support equipment will be required. These will depend on the type of job and the equipment selected, but are likely to include power units, such as electrical generators, hydraulic power units and air compressors. A three-phase mains supply offers a valuable alternative to site generation. Among other items may be a lubrication mixing and injection plant and pumps for controlling water in the shaft.

On-site accommodation has to be provided for all equipment, including the central command console from where much of the operation is remotely-controlled. As an increasing amount of pipejacking has to be undertaken in urban and suburban locations where available space is limited and noise and disruption levels restricted, planning the layout of equipment around the shaft is an important issue. The layout has to be designed to occupy a minimum area and be adapted to fit the particular site circumstances. It also must provide safe and efficient means of bringing pipes to site and transporting soil away. Containerised and stacked equipment units are now options. Some jobs require total enclosure including roofing. Enclosures with heating or cooling may also be necessary if the work is in extreme climatic conditions.

5 Equipment: microtunnelling

5.1 Classifying microtunnelling equipment

Microtunnelling equipment is relatively new on the tunnelling scene. First introduced in Japan in the 1970s, it was extensively developed during the 1980s to become established in Europe, and then in North America in the early 1990s.

The term *microtunnelling* is widely, but not universally, accepted to describe the technique of installing a pipe of less than DN 900 mm to a predetermined line and level by remote control of the head. The term encompasses all the elements of the system, not just the equipment. As the definition is based on non-man-entry working size and this varies from one country to another, 900 mm is a somewhat arbitrary demarcation. It is quite possible that, in the future, safety regulations may increase the diameter considered safe for man-entry working.

The term is sometimes loosely used for any remote-control shield equipment where the control operation is located outside the line being driven. Many manufacturers produce a range of machines which span both pipejacking and microtunnelling diameters. For example, the Iseki Unclemole is available in 10 models to install pipes in diameters ranging from 250 mm to 1800 mm. A more appropriate generic term to cover this range of remote equipment would be *remote small-bore jacking*.

As microtunnelling is a sub-category within pipejacking, a great deal of the information presented in Chapters 3 and 4 is relevant here. This is particularly true for pressure balance principles and equipment. Microtunnellers can be classified into four main groups:

- Pilot bore
- Auger-based
- Pressure balance
- Soil displacement

In Japan, all four methods are used, whereas auger and pressure-type microtunnellers have dominated the European and North American markets. Hybrid systems are also emerging which incorporate features from more than one of the main classification groups.

Excluded for consideration here are types of equipment where steering is made from the pit by mechanically turning the whole line or by some form of mechanical linkage to change the head attitude. Japanese equipment of this kind has been mainly used for installing steel sleeves.

One major difference between pipejacking and microtunnelling is that IJSs cannot be set up in non-man-entry sizes in order to extend the drive length. In practice, however, for some larger microtunnelling jobs, DN 750 mm and upwards, IJSs have been fitted. Another difference is that spans achievable by microtunnelling are less than for pipejacking. Span length is controlled by the maximum safe end load that can be applied to the pipe; the larger the diameter the larger the safe end load.

Directly installing the product pipes by jacking them up behind the cutting head as it advances is called a *one-pass* or *single-pass* installation. To reduce the load on the pipe and to drive greater lengths, the *two-pass* system is used with some microtunnelling systems. In the two-pass system a purpose-built temporary steel liner is installed first, After the bore is completed, the permanent pipe is jacked in, replacing the temporary pipe, which is recovered at the reception pit. Undoubtedly two-pass systems allow greater loads, reduce damage to the pipe and can achieve rather longer drives. But these advantages are offset by the time required for the additional operation. Two-pass systems are operated with both auger and pressure balance equipment. Theoretically any equipment can be adapted to work in this manner, but in practice a two-pass installation is an integral part of a manufacturer's system.

Microtunnelling equipment is designed and marketed in complete integrated packages. Each group will therefore include its own ancillary items, such as cutting heads, muck disposal and monitoring systems.

5.2 Pilot-bore microtunnelling

Pilot-bore equipment, the original microtunnelling method, was introduced into Japan in 1973. The principal manufacturers are still Japanese: Komatsu with the TP 80 Iron Mole and Nitto Koji with the Earth Arrow. The Iron Mole is shown in Figure 5.1.

Installation by pilot-bore microtunnelling is a two-stage operation. The first stage is to drive a pilot bore on the required line and level. This can be done in two ways: by cutting and removing the spoil or by thrust-boring which displaces the soil. In either case, progress is monitored and steering corrected accordingly. The second stage is to ream back through the pilot hole to enlarge it to the required diameter. This can be done by non-pressure reaming or by pressure reaming. Various reaming heads can be fitted for non-pressure reaming. Three types of head are available for pressure reaming according to the type of ground: a standard head and special heads for reaming in sand or gravel and for working in hard pan.

5.2.1 Stage one, the pilot bore

The pilot bore can be driven by cutting through the soil or by displacing the soil. Which technique is selected will depend on the ground conditions.

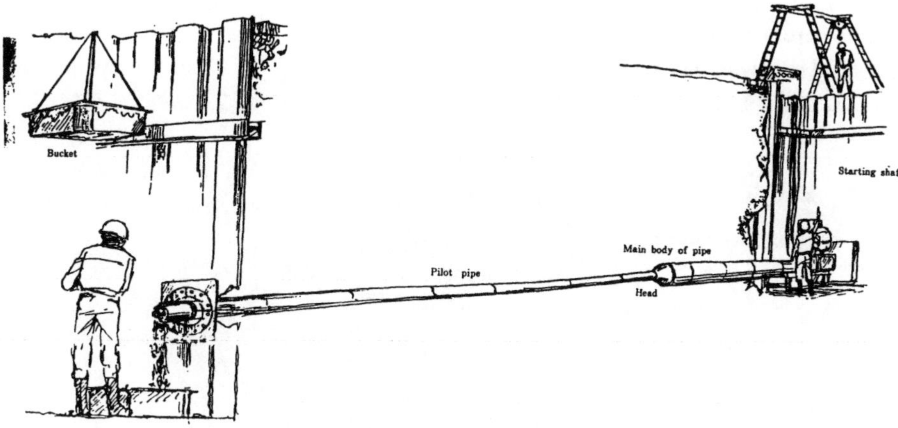

Figure 5.1 Pilot-bore method - Komatsu Iron Mole (sketch by courtesy of Japanese Pipe Jacking Association)

Cutting. Rotating cutter heads are needed to cut through harder soils and granular materials. Various heads are available to suit different conditions. As an alternative pilot-boring device, the Iron Mole has a rotary cutting head fed with water. The spoil is removed back to the drive pit as a slurry. Also available are cutting heads for working in harder material and a crushing head for sands and gravels, as seen in Figure 5.2 The alternative pilot system for the Earth Arrow (Figure 5.3) is designed for driving in firmer soils. It has a chisel head, and conveyor screws for removing the cuttings.

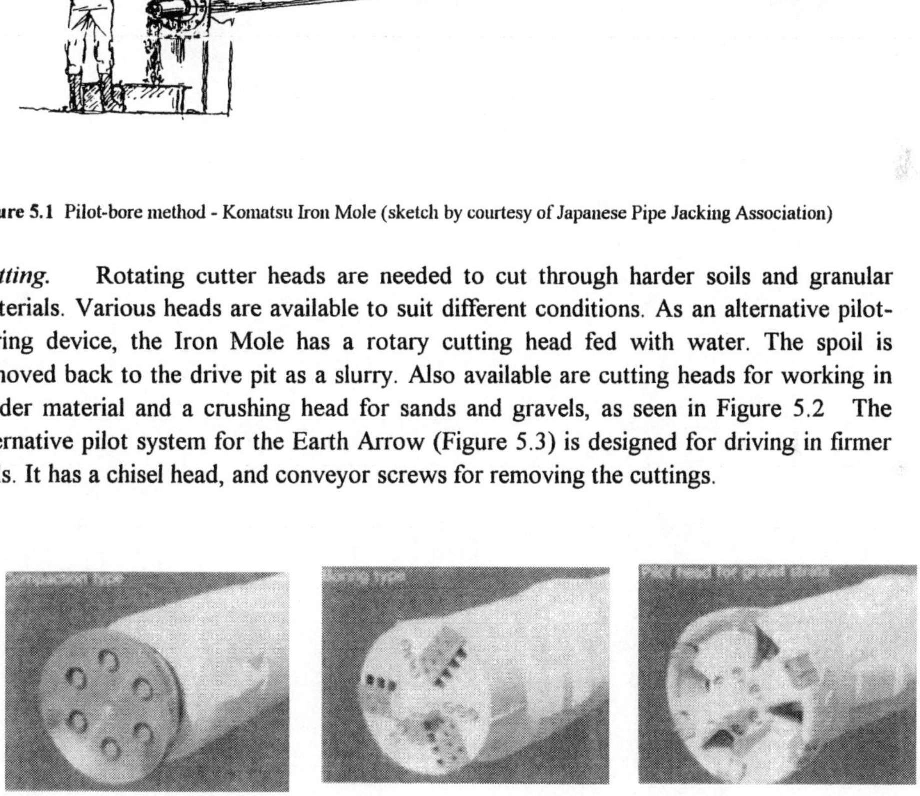

Figure 5.2 Iron Mole - alternative pilot heads (illustration by courtesy of Komatsu)

Displacement by thrust-boring. With this technique the soil is not cut or removed but displaced by consolidation. For this, the Earth Arrow uses a rotating solid chisel-shaped head as shown in Figure 5.4. The head is rotated for normal progress, but, when steering corrections are required, the head is oriented and pushed forward without rotation. A deflecting reaction from the soil pushes the chisel head in the required direction. The Komatsu TP 80 Iron Mole has a non-rotating anvil-type head, powered by a 50 tonne jack with a 400 mm stroke, to form and consolidate the pilot bore. Figure 5.5 is an example of this type of head and the monitoring and control system used for steering.

Displacement heads are for working in softer soils, where consolidation is possible without creating problems associated with soil movement. Suitable soils are cohesive soils, such as clays and silts with N values (SPTs) up to 20. For sandy silts, the N value drops to around 12. Diameters of displacement pilot bores range from 150 mm to 220 mm.

Figure 5.3 Earth Arrow system (Photograph by courtesy of Nitto Koji Co. Ltd.)

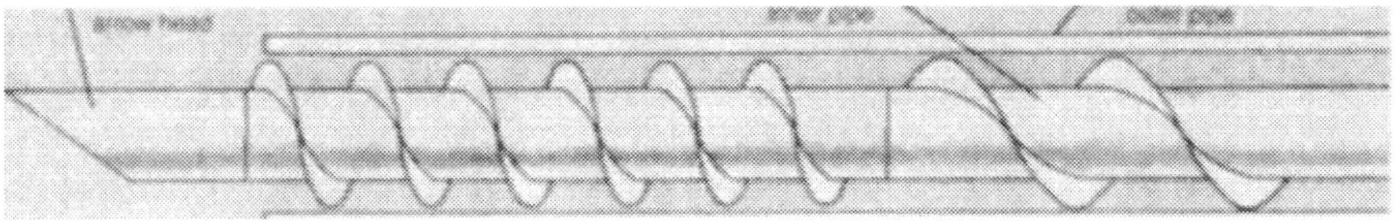

Figure 5.4 Earth Arrow heads (illustration by courtesy of Nitto Koji Co. Ltd.)

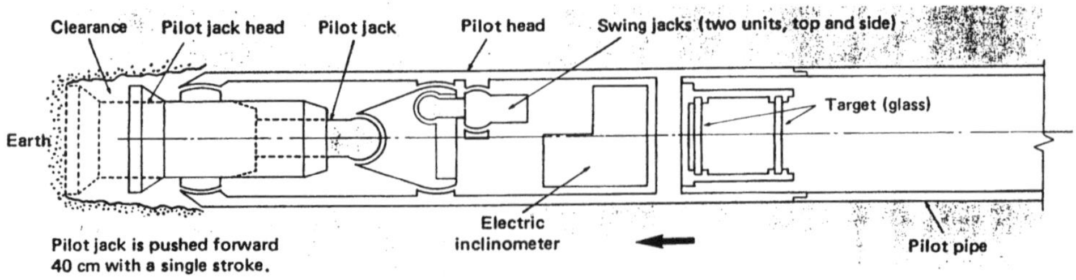

Figure 5.5 Anvil head pilot borer - Komatsu Iron Mole (diagram by courtesy of Komatsu)

5.2.2 Stage two, enlarging the bore

The bore is enlarged to the required diameter with a rotating reamer which cuts the soil and forces it into the reaming chamber, where it is carried away either by flight augers or as a slurry. The design and choice of reamer heads closely follows standard microtunnelling shield principles. In stable soils, such as clays, the cutter head, equipped with bits and teeth, cuts and forces the material into the chamber. In non-cohesive soil and where there is ground water pressure, the chamber is pressurised with water and the reamer face is of the closed type, with slots similar to those of slurry shield layouts. Other special heads are available for working in more compact granular material. Alternative reaming cutters are shown in Figure 5.6 and the operation of the reamer in Figure 5.7. The product pipe is jacked in behind the enlarging head. Table 5.1 sets out the manufacturers' combinations of heads and reamers for differing soil conditions. Table 5.2 lists the recommended range of applications.

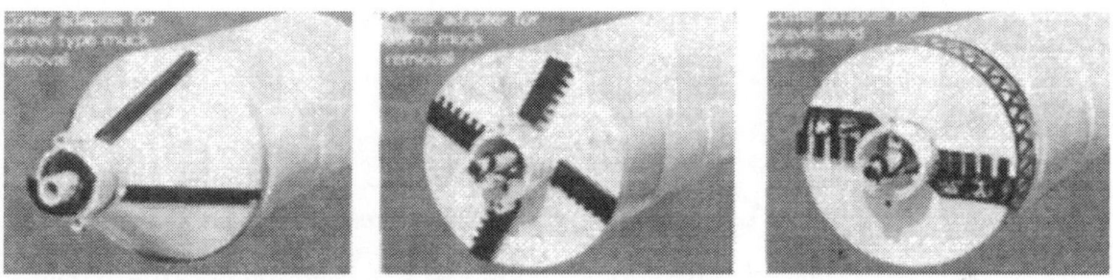

Figure 5.6 Iron Mole -alternative reaming cutters

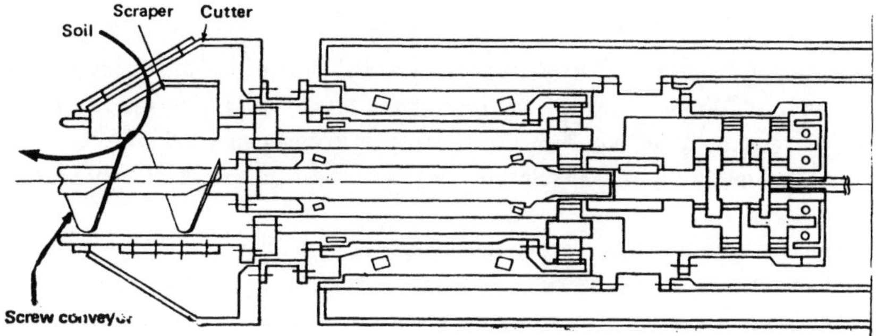

Figure 5.7 Iron Mole - reaming device arrangement

Table 5.1 Combination of heads and reamers for different soil conditions (after Komatsu)

Method	Combination		Main Features
	Pilot head	Reamer Head	
I	Displacement	Screw discharge	Most common combination especially in soft soil.
II	Cutter discharge	Slurry	Applicable to harder denser soils and below ground water table. Soil treatment only necessary adjacent to shafts.
III	Cutter discharge	Screw	Not suitable below water table. Suitable for hard soils with SPT values up to 50.
IV	Displacement	Slurry discharge	Suitable for soft and water-logged soils. Soil treatment only necessary adjacent to shafts

Table 5.2 Range of soil types for the Iron Mole (after Komatsu)

Soil conditions	Soil quality	Clay		Silt		Sand/sandy soil			
	N Value	0 ~ 20	20 ~ 50	0 ~ 20	20 ~ 50	0 ~ 13		13 ~ 50	
	Hydraulic pressure (PSI)	–	–	–	–	0 ~ 0.4 (5.7)	0.4 ~ 0.6 (5.7 ~ 8.5)	0 ~ 0.4 (5.7)	0.4 ~ 0.6 (5.7 ~ 8.5)
Method	I	◎	X	◎	X	◎	X	X	X
	II	*O	O	*O	X	*O	*O	O	◎
	III	*O	◎	*O	◎	*O	X	◎	X
	IV	O	X	O	X	O	◎	X	X

◎: Recommended O: Possible *O: Suitable when N value is more than 5. X: Not suitable.

Pilot bore methods have been used to install product pipe of various kinds in diameters from 200 mm to 700 mm. Claimed driving accuracies are ±30 mm on grade and ±50 mm on line. A maximum jacking distance of 50 m is recommended. However, pilot-bore microtunnelling has not been generally adopted outside Japan for a number of reasons:

- Types of soil in which it works successfully are limited.
- The pilot bore can be difficult to control as it is relatively easily deflected.
- The longer the drive, the greater the chance of deviation.
- When a deviated bore has been corrected, pilot bore and product pipe may be subject to high stress.

Forms of pilot bore equipment are discussed in Section 5.4, on house connections.

5.3 Auger-bore microtunnelling

In its simplest form, auger-bore microtunnellers incorporate augers (screw conveyors) which remove the soil from the bore and also power the cutting head. Although developed from it, auger-bore microtunnellers differ from traditional auger-boring in so far as the articulated two-part head can be monitored and steered. A second difference is that the screw conveyors operate inside a steel casing contained within the sections jacked in behind the head.

Auger-bore microtunnellers, which are designed with direct drives to the head or with an additional independent drive for the head, can all be fitted with a range of head variations. Installation can be by one or two pass methods. Some auger-bore microtunnellers are designed to install plastic pipe, which cannot be done by jacking.

5.3.1 *Drive types*

Direct drive. In direct-drive models, the screw conveyor provides the soil removal system and also serves as the shaft to power the head. The equipment shown in Figure 5.8 has the drive, jacking, and operator control all within the same power unit, which moves forward on the sub-frame as the pipe is pushed through the pit. After each section of pipe is jacked through the pit, the drive is disconnected and the whole unit moved back. A new section of pipe, with the auger within its steel casing, is dropped into position, connected and the pipe joint made.

The conveyor screws discharge the soil back into the starting pit. The auger is contained within a casing firmly located on the centre line of the pipe being installed. Cutting head and augers turn at the same speed, which can be regulated from zero to (in some machines) more than a 100 rpm.

Figure 5.8 Horizonger auger microtunneller (photograph by courtesy of Sanwa Kizai Co. Ltd.)

The cutting head rotates inside the articulated shield, which is fitted with rams for remote angular adjustment and also a target for laser-beam or theodolite sighting. The space between the internal diameter of the pipe and the outside of the auger casing provides a path for the monitoring. Figure 5.9 shows this arrangement. Theodolites and laser theodolites are used for observing targets mounted in the boring head. A prism device is often used to facilitate a line of sight down the pipe. Some auger-driven machines have hollow stem flights to allow theodolite observation of a Light Emitting Diode (LED) target mounted in the head.

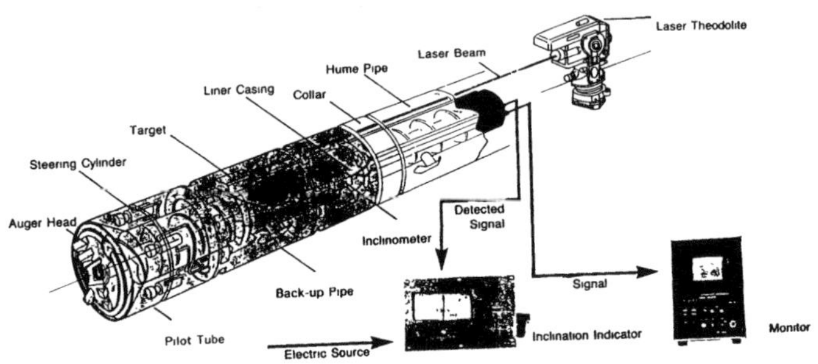

Figure 5.9 Horizonger monitoring system (illustration by courtesy of Sanwa Kizai Co. Ltd.)

In the original Japanese versions of this type of machine, the main power unit and controls were located in the working pit, which meant having a drive shaft of 5 m or more. Subsequent European versions separated the power generation unit from controls, situating these outside the working pit. With a compact jacking system and only the drive motor for the screw conveyors in the drive shaft, the length of the drive shaft can be reduced to 2 or 3 m using short-length pipe (1000-1200 mm sections). This type of equipment is shown working from a circular shaft in Figure 5.10. Auger-bore microtunnellers of this basic type have proved to be particularly suited to driving smaller diameter pipes over spans up to 60 m in dry non-cohesive soils. Where diameters get greater and spans longer, the torque available for the head and driving the screw conveyor may be inadequate. Problems of auger-flight coupling failure have also been noted.

Direct geared-drive type. To overcome the limitations of direct-drive auger microtunnellers, some manufacturers have introduced a gearbox within the drive head connecting the main line auger and the head drive. The gearbox ratio is set at around 4:1 so that the head is rotating at about a quarter of the speed of the augers but with a much greater torque. This extends the machine's capability quite considerably.

Independent drive type: A logical development for extending capability and flexibility still further was to provide an independent drive for the cutting head. Material from the head is discharged into a chamber where it is picked up by the screw conveyor system. Figure 5.11 illustrates this type of head. With such an arrangement it is possible to install 600 mm pipes over 100 m or more. Independent drive heads can also work more efficiently in tougher soil conditions, including soft to medium hard rock.

An example of this type of head is used in Soltau's RVS 300, which can be adapted to the following modes:

- As an auger-driven head for soft to medium-hard ground above the water table.
- As a separate drive head with separately driven spoil-removal augers. The head can operate as a crusher, so is suited to tougher soil conditions.
- As a direct-drive excavation head with spoil removal by slurry for difficult soils below the water table. (This mode is an example of a hybrid design).

Figure 5.10 Small circular shaft for auger microtunnelling

5.3.2 Head variations

Types. According to ground conditions, a range of cutting heads can be fitted to auger-bore microtunnellers. The simplest forms are based on the flight auger and have a spiral cutting head. Picks, chisels and bits can be mounted on the wings to improve cutting capability. A head of this type is illustrated in Figure 5.12. With direct-drive machines it is sometimes possible to withdraw the auger string and head in order to change worn cutters or fit a different type of head to meet changed ground conditions.

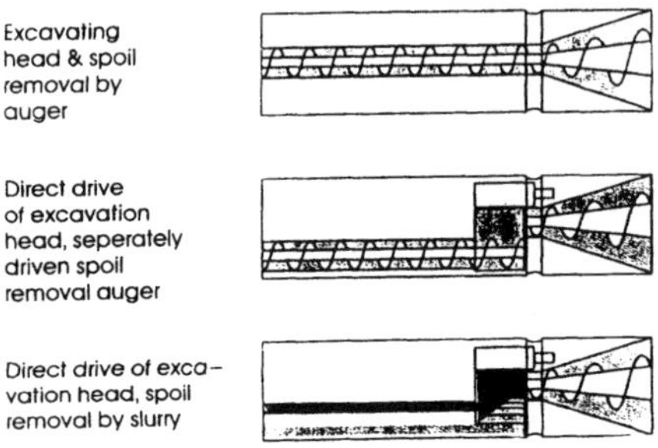

Figure 5.11 Independent drive head
(drawing by courtesy of Dr. Ing. G. Soltau GmbH)

Figure 5.12 Winged auger head

Heads for heavier duty can be spoked, flat-plated or dome-shaped. Scrapers, bits, chisels, disc and roller cutters can all be fitted. A spoke head from Soltau is shown in Figure 5.13 and a dome-head from Komatsu TP 90 in Figure 5.14. Cohesive soils can cause clogging of the head. Water injection and cleaning rods are built in to some heads to deal with this.

Figure 5.13 Spoke head for RVS 250H
(illustration by courtesy of Dr. Ing. G. Soltau GmbH)

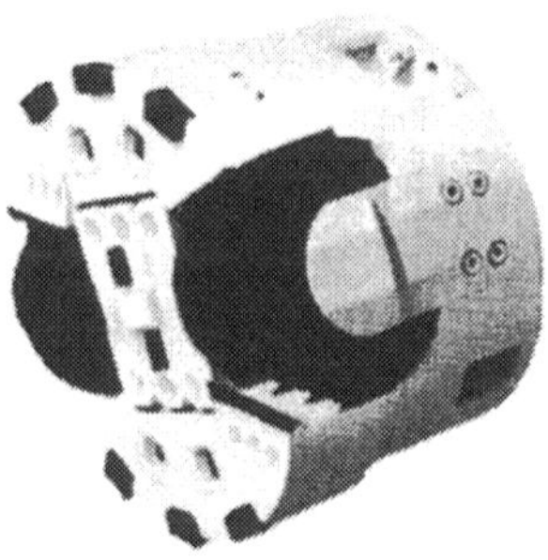

Figure 5.14 TP 90 disc cutter head (photograph by courtesy of Komatsu)

The Super Striker method from Japan incorporates a multiple-head down-the-hole hammer within a coring head. Normal progress is by coring, a percussion device being used to break up obstructions and rock. A perimeter cutting head (trepanner), shown in Figure 5.15 is combined with a an auger type of head on a machine from Witte of Germany. The cutting rim can be rotated counter to the rotation of the auger head. Figure 5.16 shows a crusher head, from Sanwa in Japan, which rotates within a shield lined with a series of hardened cam profiles. As the material is forced back through the narrowing cam aperture, it is crushed to a workable size for removal by the flight augers.

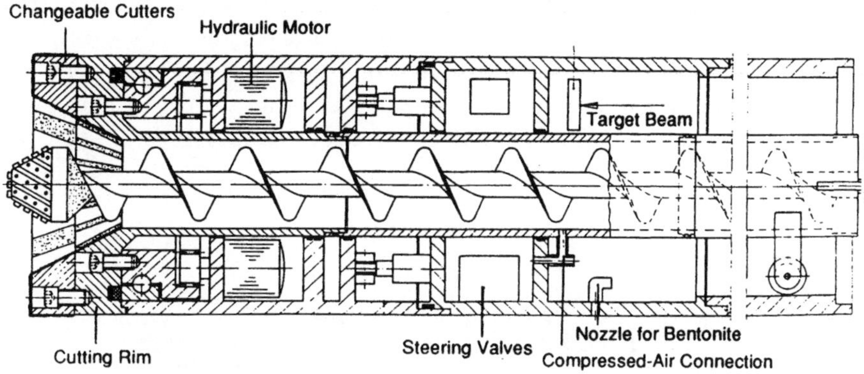

Figure 5.15 VM-F400 trepanner head (drawings by courtesy of Witte Bohrtechnik)

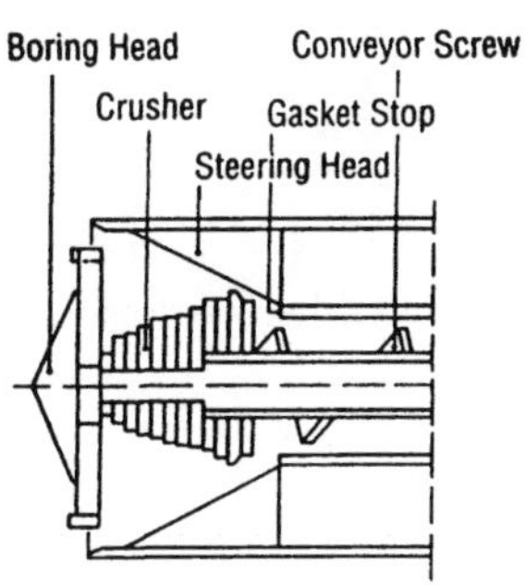

Figure 5.16 Presstone system crusher head (drawing by courtesy of Sanwa Kizai Co. Ltd.)

Pressure balance. An auger system is equivalent to an open shield in pipejacking. To work safely below the water table requires some form of pressure balance device and a closed system. Some machines have the option of injecting pressurised air, water or slurry into the head to provide pressure balance. The ways in which the system is pressurised vary from one manufacturer to another, but all serve to seal either the cutter head or the whole auger system. Figure 5.17 shows the layout for sealing the whole auger system, with a water pressure chamber on the end of the line. A mechanical earth pressure balance can be achieved by fitting a gate valve or rotary vane feeder on the end of the auger string. However, this type of device has limited capability in working below the water table (around 3 m). They can also complicate and slow down progress.

The independent drive head discussed above incorporates a transfer chamber at the rear of the cutting head. This chamber can be relatively easily pressurised by a slide valve between crushing head and chamber, allowing work below the water table.

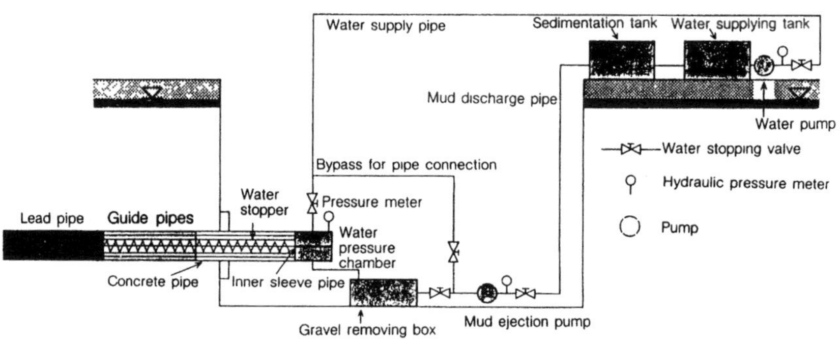

Figure 5.17 Pressurisation of an auger line - Smole system (diagram by courtesy of the Smole Association)

Spoil removal The larger the auger flight the greater the torque that can be transmitted and the larger the grain size of material that can be conveyed. The largest conveyable particle size is about 75 mm.

There is an optimum diameter for auger and casing in relation to the pipe being installed. However, the costs of carrying a wide range of screw conveyors and casings makes this impractical. Standard augers working inside the steel casing are suited to removing non-cohesive materials such as sands. Large gravels and stiff clays can cause blockage problems, although modified flights are available which can cope with these types of spoil. Where transfer points are used (inevitable at the pit bottom), there will be spillage problems.

The Pezzimenti Laserbore microtunneller has been developed by an Australian contractor and operated by his own crews on local work. A unique feature of this rig has been its vacuum system for spoil removal. A truck-mounted vacuum system is connected by 120 mm steel ducts to the auger-type boring head. Pezzimenti claim to handle cobbles up to 80 mm and, so long as it is dry, to extract soil directly. Water is added to moist soils to make them flow more easily and avoid clogging, while excess water is drained away from the vacuum truck before disposal [26]. KCMM, a Japanese company, also use a vacuum extraction system with an auger cutting head.

The cassette method developed by Westfalia Becorit of Germany is an ingenious attempt to overcome some of the limitations of auger machines. Figure 5.18 illustrates the operation of the WBM-C10. A cassette is fitted with a drive motor which drives the cutting head and the short length of screw conveyor in the container. In the driving position, the cassette is docked into the drive head and excavation from the head is packed into the container. When full, the cassette is undocked and pulled back to the working shaft, where it is emptied and returned to the face. It is also possible to extract the head with the container and replace or renew cutting bits. The drawback is that, because of the intermittent nature of the operation, drivage is reduced to about 6 m per shift.

5.3.3 Line installation

Two-stage installation. A number of auger machines use two stage installation. In Japan, Toho and The Smole Association have methods which initially install a temporary double steel wall casing. The temporary casings in the Smole method are shown in Figure 5.19. A Witte auger bore machine used a similar method. Several house connection machines also employ two-stage working.

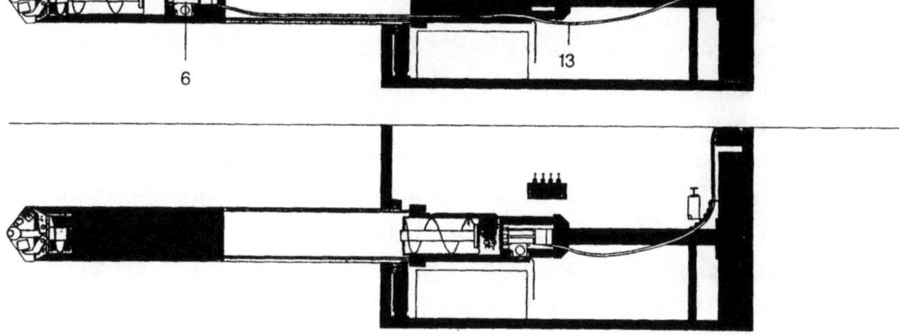

Figure 5.18 WBM-C10 cassette system (drawing by courtesy of Westfalia Becorit)

Figure 5.19 Smole two-pass temporary casing (illustration by courtesy of The Smole Association)

Plastic pipe installation. Although used for small diameter sewer installations, plastic pipes are not suited to jacking as they have poor end loading capability.

The Enviliner system from Japan is a basic auger machine but has screw conveyors working inside a casing, which is itself situated inside the PVC pipe. Thrust is transmitted down the temporary casing on to the shield; not through the PVC pipe. Diameters up to DN 300 mm are installed by this method.

The LLB method takes the concept further by introducing *grippers* [27]. The thrust is again transmitted down temporary casings, but thrust force is transmitted onto the permanent plastic pipe by way of air-powered gripper mechanisms mounted at appropriate intervals on the casing, as seen in Figure 5.20. An LLB gripper system was tested at Louisiana Tech University and the results are described by Najafi [28].

5.3.4 Equipment manufacturers and suppliers

Worldwide there are around 20 manufacturers of auger type microtunnelling equipment, each having particular features and variations on the basic method. Machines are available normally in a range of four sizes, covering diameters 150-900 mm. Modifications within each of the four sizes can be made by fitting external skin plates to the exterior of the shield. So adapted, one machine can drive pipe diameters from 250 to 400 mm.

Japanese manufacturers of auger-bore microtunnellers have their main markets in Japan and some other South Eastern Asian countries. However, the Sanwa Kizai Horizonger range, and particularly their adaptations for installing plastic pipes, have created interest outside Japan. Komatsu, although an international supplier of construction equipment, has not yet promoted its auger microtunneller outside that country.

In Europe and in North America the leading designer and supplier of auger-based machines is Dr. Ing. Soltau of Germany, who, in an eight-year period from 1984 has sold more than 150 machines. Other European manufacturers of auger equipment include the German companies Witte and Westfalia Becorit, and Dosco of the UK. In North America, Soltau has its own subsidiary. The American Auger Company, with long experience in auger-boring equipment, has announced its intention to market its own design of auger-based microtunnellers.

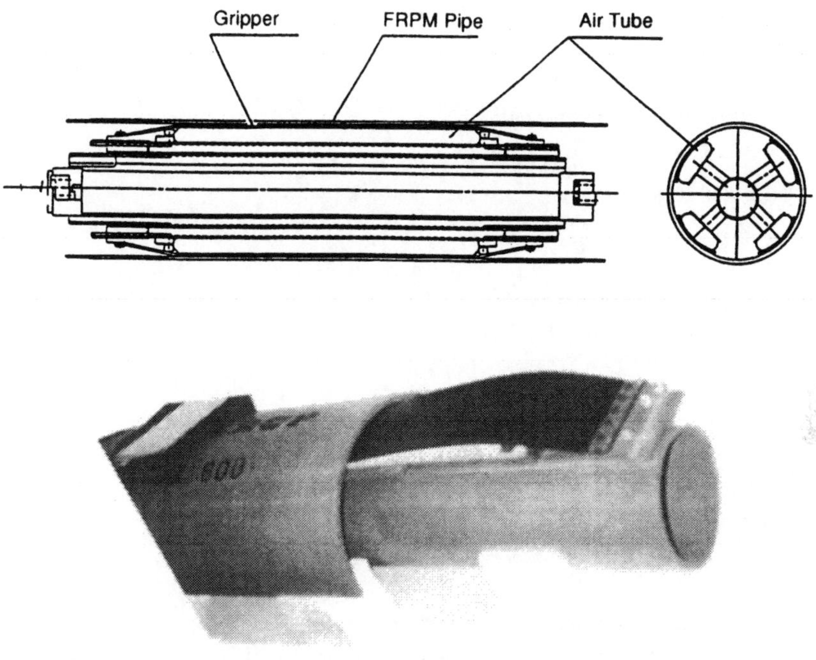

Figure 5.20 Pipe Gripper mechanism (source SKK System)

5.4 Pressure balance microtunnelling equipment

Pressure balance microtunnellers, probably the most widely used type, are miniaturised versions of the pipejacking pressure equipment described in Chapter 3. They are characterised by being a closed system with full-face shields, a pressure chamber with a supporting fluid, and continuous slurry spoil removal. The first pressure balance microtunnellers were developed in Japan in the late 1970s. In the 15-year period up to the end of 1992, an estimated 2000 to 3000 machines had been sold. The original machines required shafts of 6 and 7 m in length. Most present equipment, working with short-length pipe sections of 1000 to 1200 mm, can be accommodated in circular shafts of 2500 to 3000 mm in diameter.

Pressure balance microtunnellers can be divided into three groups according to the type of pressure used, water, bentonite slurry or earth-balance. Table 5.3 shows the type of head that can be used with the first two types of pressure balance microtunnellers. Earth balance machines have their own specific heads.

Table 5.3 A classification of pressure balance microtunnelling equipment

Type according to pressure	Type of head
Water	Standard Eccentric cone Crusher
Bentonite slurry	Standard Crusher
Earth balance	(Specific heads)

5.4.1 Water balance machines

Most microtunnellers use water as the fluid medium with bentonite and polymer muds added when conditions are very unstable. Machines are normally capable of working with either water- or bentonite-based muds.

Eccentric cone. Iseki's Unclemole, perhaps the most widely used pressure balance microtunneller, is one that operates with water. Several models of the Unclemole, for which *skinning up* facilities are available, cover pipe diameters from 250 mm up to 900 mm. The eccentric cone head and shield can be adapted to drive larger diameters by using external skin plates and a larger cutting head. The Unclemole can work in soil conditions ranging from N values from 5 to 50, and cope with ground water heads of up to 10 m. All excavated material is crushed in the head stones, including cobbles up to 30% of the shield diameter. Installation lengths of up to 180 m are possible for 600 mm pipe.

The Rockmole, a variation of the Unclemole with increased power and a more robust construction, is designed for work in rock with strengths up to 100 N/mm^2. A cutter head is mounted in front of the cone crusher. The head, shown in Figure 5.21, has two types of roller bits for cutting the face and drag teeth for directing material into the crushing chamber. Machines of this type are available for installing pipe diameters from 600 mm upwards.

Crusher chamber. Machines with crushing heads are available in microtunnelling sizes. Herrenknecht's AVN series, a machine used extensively in Europe and the USA, has a water-based pressure balance for ground water heads up to 15 m. Of the 10 models in the range, 6 fall into the microtunnelling category. The smallest diameter that can be installed is 250 mm. Variations in outside diameter can be achieved by skinning up the shield and increasing the diameter of the cutting head. Depending on pipe size and ground conditions, drive lengths average between 100-150 m, but greater lengths have been installed.

Figure 5.21 Cutter and drag teeth on Rockmole (photograph by courtesy of Euro-Iseki Ltd.)

All AVN machines have behind the cutting head a crushing chamber which breaks down cobbles and rocks into fragments small enough to be pumped. Figure 5.22 shows this arrangement. Various head configurations are designed to deal with different soils. Figure 5.23 shows standard heads and rock heads. Even standard heads are equipped with carbide-tipped cutters, which can cut through mass concrete entry and exit walls.

The Iseki Crunching Mole for microtunnelling comes in three models for pipe diameters of 600 mm and greater. The action is the same as that of the larger pipejacking equipment, with a rotary crushing chamber. The maximum size of cobble that can be dealt with by the 600 model is 170 mm.

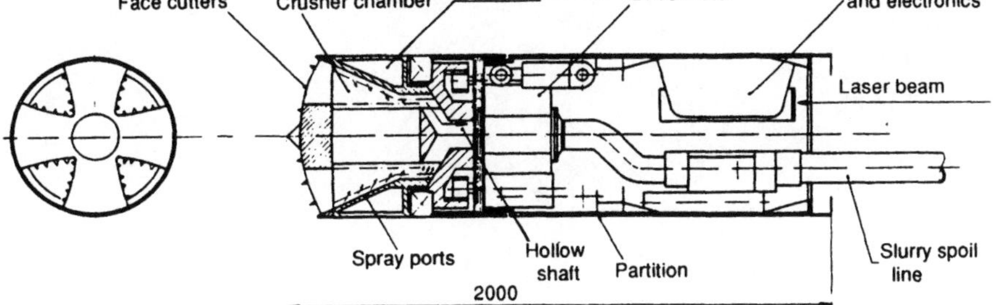

Figure 5.22 AVN crusher head (drawing by courtesy of Herrenknecht International Ltd.)

Figure 5.23 AVN water pressure balance microtunnellers - standard rock heads (by courtesy of Herrenknecht International Ltd.)

5.4.2 Bentonite slurry machines

Rasa pressure balance equipment, much used in Japan, comes in several types, including cone crushing and jaw crushing versions. Each of these is produced in microtunnelling sizes. Bentonite- and polymer-based slurries are generally used in Rasa equipment. Work can be undertaken in up to a 30 m water table head.

The Okumura Super-Mini operates with either water or bentonite-based muds and is an example of a two-stage system. The slurry-pressure balance head shown in Figure 5.24 can deal with a range of soil conditions and with ground water heads of up to 25 m. The temporary pipes used at the first stage of installation contain all the slurry and service pipes; connections are made and sealed automatically as each new section is set in the drive pit. The cables, carrying power, control and communications, are continuous and are laid in a slot on the top of the casing and then protected with a steel cover. Purpose-made sections of this type can be quickly installed. As they are bolted together, it is possible to retract the head and line of sections in the case of an aborted drive. Should this be necessary, it is possible to backfill the void with pressure-injected material to avoid any settlement. Drive lengths up to 200 m have been completed.

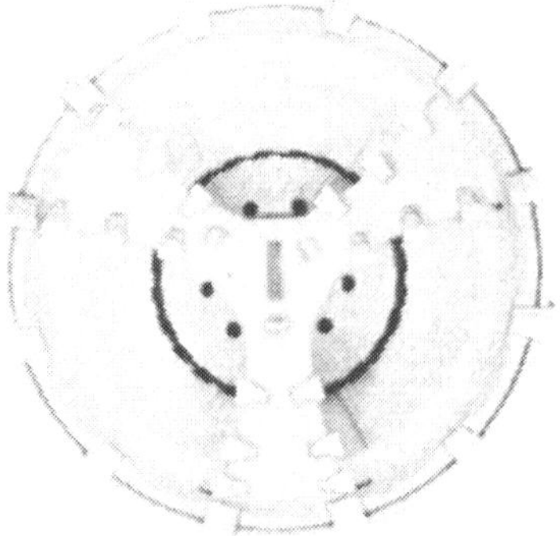

Figure 5.24 Super-mini slurry pressure balance head (illustration by courtesy of Miller-Markham Ltd.)

Temporary casings also allow the introduction of a reamer unit, which can be used to enlarge a bore from, say, the basic 500 mm to any size up to 900 mm. The temporary pipes are installed in the normal fashion and the shield head recovered at the reception pit. The reamer unit is then connected to the end of the temporary pipes and the permanent pipes to be installed are fed in behind the reamer. A second jacking rig set up in the reception pit can pull back the permanent pipes and reamer, while the lengths of temporary pipe are recovered in the start pit. The rotating reamer operates exactly like a shield, having cutting teeth and pressure balance, with slurry transport of material down the temporary steel pipe [29]. The reamer device is shown in Figure 5.25.

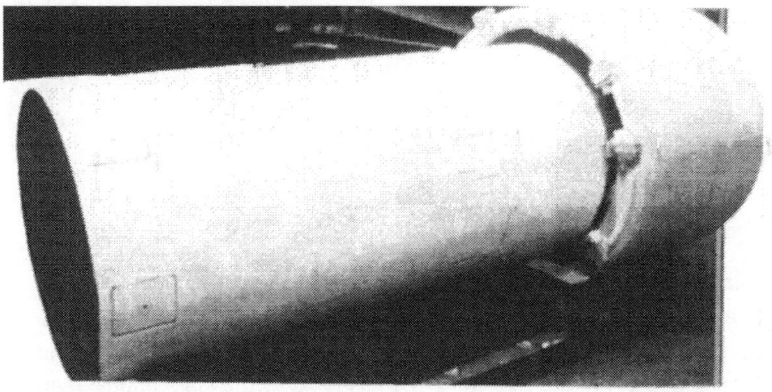

Figure 5.25 Reaming device for Super-Mini (photograph by courtesy of Miller-Markham Ltd.)

5.4.3 Earth pressure balance machines

Originally known as the Telemole, the Iseki microtunnelling earth pressure balance model is now classified under the general MEPCB name. Three microtunnelling models are produced for pipe diameters of 600, 700 and 800 mm. As for the larger pipejacking machines, the earth balance is achieved by limiting the amount of spoil that passes through the adjustable face apertures and the load applied to the soil face by the head. The most suitable soil conditions are silts and sands with N values less than 30 and saturated clays with N values up to 15. The maximum particle size that can be handled is 50 mm. The MEPCB uses a water-based pumped slurry removal system.

An unusual microtunnelling machine known as the Ace Mole DL35, Figure 5.26, has been developed by Nippon Telegraph and Telephone (NTT) of Japan. An interesting feature is that bentonite, polymer lubricant and plasticisers are injected into the soil at the face and mixed by a head rotating at 50 rpm to produce an area of *liquefied soil* around the head.

No muck inlets are in the face, but as the machine advances through the slurried area, material is recovered through a port located towards the rear and at the bottom of the shield. This microtunneller has installed several kilometres of pipe in diameters from 350 to 600 mm. Its greatest use has been in clays and silts with N values less than 30.

Figure 5.26 Ace Mole DL35 (photograph by courtesy of NTT)

5.4.4 Manufacturers and suppliers

Japanese manufacturers have been a dominant force in the pressure balance microtunnelling market, with Iseki, Rasa and Okumura as leading suppliers. About 12 others focus on the home market. Iseki has concentrated on this type of equipment and through sister companies in the UK and USA has become the international market leader. Rasa and Okumura designs are available through licensees in UK.

In Europe Herrenknecht slurry machines compete head-on with Iseki for leadership. There are several other European manufacturers, including MGF and Westfalia Becorit of Germany.

5.5 Displacement methods of microtunnelling

NTT and Komatsu offer guided thrust-boring systems which displace soil rather than cutting and removing it from the bore [30]. Section 5.2.2, on pilot-bore installation, referred to the action of the Iron Mole displacement head, used for installing ducts, including curved, in soft soils in Japan (N values up to 10). However, these are not true microtunnelling machines and therefore fall outside the scope of this book

A displacement tool, the Perimole, recently developed by Iseki does, however, meet the criteria for microtunnelling methods. It is designed for installing pipes with diameters of 300 mm and less. The head, shown separately in Figure 5.27, comprises a split cone, the top and bottom of which move in an opposite epicyclical action. This action is designed to move soil and compress it around the head. As no spoil is generated there is no disposal system. Monitoring is by target and laser; steering is by pivot rams in the head. Accuracies within 25 mm on line and level are claimed, and drive lengths of more than 50m. The method is designed for smaller diameter work in soils amenable to compaction. With displacement methods of this kind, consideration must be given to outward soil movements which may cause heave and movement.

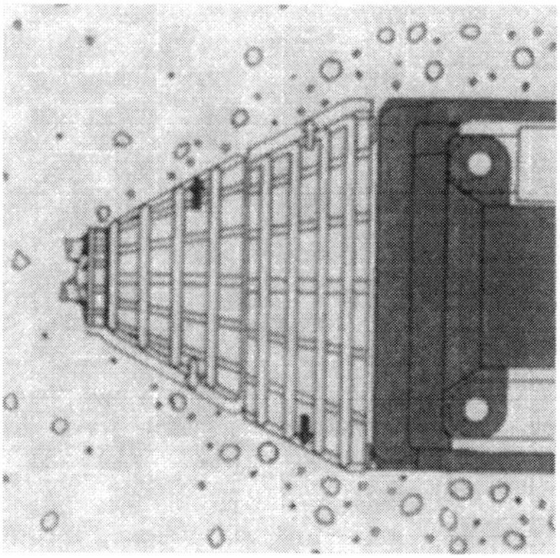

Figure 5.27 Perimole soil displacement head (drawing by courtesy of Euro-Iseki Ltd.)

5.6 Microtunnellers for on-line replacement

One of the growing demands made of trenchless technology these days is to replace ageing sewers, particularly in urban centres. Many ingenious techniques, unrelated to pipejacking and microtunnelling, have been designed to remove and replace these in a single operation, *on-line*. However, manufacturers of microtunnellers in Japan and Germany have adapted their machines for this purpose. These [32] [33] like other methods of on-line replacement, have to meet the following criteria:

- Simultaneously remove the old pipe and install a new one.
- Provide a pipe of increased diameter.
- Install to a corrected invert level or line even where the old line has deviated.
- Replace over manhole lengths.
- Work in all soil conditions and below the water table.
- Maintain flow during working.*
- Remove all kinds of existing pipe materials including reinforced concrete.*

*Some equipment on the market does not include these facilities.

Adapted from the Unclemole, the Iseki Piranha, shown in Figure 5.28, is fitted with a modified cutter head, which incorporates *clippers* for cutting reinforcing bars. Also added is a proboscis, which is entered into the pipe to be removed. This serves both as a seal to stop loss of slurry from the shield and to pick up and pump through and away existing flows in the sewer.

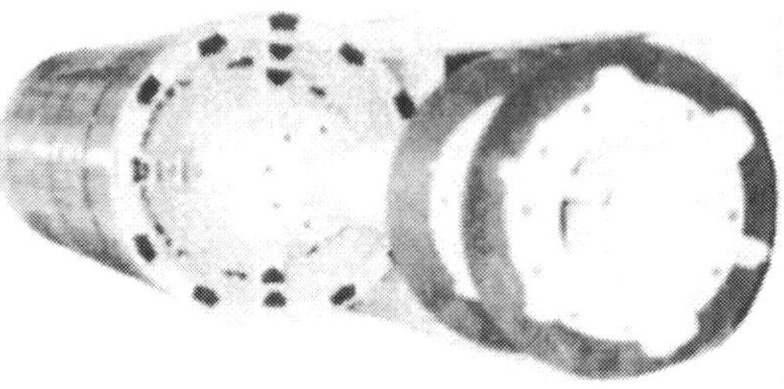

Figure 5.28 Piranha - pipe replacer (photograph by courtesy of Euro-Iseki Ltd.)

The German company, NLW Fordertechnik, has produced a purpose-built package of equipment for on-line sewer replacement. Based on an auger-type microtunnelling system, the head is of open design with a nose-cone and twin arms mounted with roller bits and hard-faced picks. An eccentric crusher breaks down all rubble into small particles, which are removed either by vacuum extraction or a screw conveyor. It can install pipe diameters of 400 to 600 mm.

5.7 Microtunnellers for house connections

The need to install not only sewers without disruption but also their connections to properties has stimulated the design of purpose-built equipment. As installation lengths are short and greater tolerance is allowed on line and level, non-steerable auger-type machines have often been used to install a temporary casing in a two-stage process. Steerable machines for this purpose are essentially simple versions of microtunnellers. Installed pipe sizes from 100 to 200 mm require a two-pass method of working. The need for pressure balance to counteract water table is seldom necessary as connections tend to be laid at shallower depths.

Equipment needs to:

- Be designed for maximum bore lengths of 30 to 40 m.
- Work from a minimum working area (typically a 2000 mm diameter shaft).
- Be easily transportable, as it is not unusual to have to work from house basements or large diameter sewers.

The Witte house connection system is a scaled-down version of the company's larger auger microtunnellers. Figure 5.29 shows the head arrangement. A theodolite in the working pit and a passive target in the head are used for monitoring. Steel casings are installed at the first pass and pushed out with the permanent pipe at the second stage.

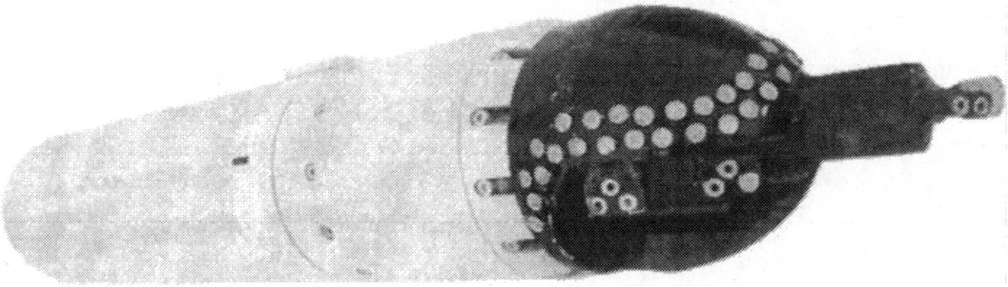

Figure 5.29 Steerable BSK head for Witte house connection method (photograph by courtesy of Witte Bohrtechnik)

The Soltau RV 35 uses the pilot-bore technique with a displacement head for the first stage. Accuracy of line and level is monitored by a camera theodolite sighted through the hollow pilot tubes onto a target at the head. An operator controls the steering based on the picture transmitted to his screen by orienting the bevelled steering head. Once the pilot line is completed to the reception shaft, an excavating reaming head is attached to the pilot string and steel casings with auger flights are pushed in behind the head as it advances. At the third stage the permanent pipes are pushed in to replace the temporary casings. Where the auger flights can be run inside the permanent pipe, it is possible to combine stages two and three.

Several other German and Japanese manufacturers have similar auger and pilot-bore microtunnelling systems for house connections. All these two-pass methods need to work from a drive pit and have a target pit for recovering sections and for making the final connection.

A German manufacturer, Bohrtec, has specialised in house connection equipment. Though similar to the methods already described, these machines have been designed so they can also operate from within large-diameter sewers. Bohrtec's sophisticated HABM system allows connections to be made remotely, particularly valuable when a property is being connected into a deep sewer located in the street. A steel casing is installed, which is fixed at the joints in the horizontal direction, but with flexibility in the vertical direction. Steering for gradient is therefore possible. An auger-powered cutting head inside the casing cuts and removes the soil. On reaching the existing sewer, the augers are withdrawn and a diamond drilling bit inserted to core the sewer wall. A special sealing element is attached at the front of the pipes to be installed and the string of pipes is pushed into the casing. The sealing unit mates into the cored hole to provide a watertight joint. The outer casing can then be retracted and the annular space filled with injected material [34]. This is a relatively high cost procedure which is only warranted in a few situations.

5.8 Site equipment

Manufacturers are designing equipment with compactness in mind. Microtunnellers have been reduced to a size where they can work from shafts of 2 to 3 m in diameter, so causing minimum disruption to traffic and public. Auger-bore methods require less room than the slurry pressure-balance methods with their mud separation plants. Both auger-bore and pressure balance microtunnelling equipment is now often packaged in standard ISO freight containers. The container is placed over the head of the drive shaft to give full protection against the weather. It is fitted with an extendible overhead crane beam for handling muck skips and pipes, and also houses the operator's control panel and the power units. Figure 5.30 shows the interior of a container and Figure 5.31 gives a clear impression of the compact nature of working. However, the most recent trend is to reduce the container size to accommodate just the equipment and controls and not to set it over the shaft.

Figure 5.30 Container interior (photograph by courtesy of Dr. Ing. Soltau GmbH)

Figure 5.31 Site with containerised working (photograph by courtesy of Dr. Ing. Soltau GmbH)

6 Design concepts

6.1 Role of client, engineer and contractor

The roles of the three traditional parties to a contract for infrastructure works are changing. This chapter discusses some of the reasons for these changes and identifies some of the trends of change evident in different parts of the world.

6.1.1 Traditional approaches to design

Traditionally the client was a national or local government authority with monopolistic status. Policies and investment were often ruled by political expediency and had little bearing on the public good. Consulting engineers, on fixed scales of fees, provided engineering designs for the capital works and were considered to adopt an independent position between client and contractor. The contractor, essentially a supplier of men, materials and machines, followed the instructions of the engineer through plans and documents. None of the parties had much incentive to innovate or even to reduce costs. This three-cornered arrangement often provided fertile ground for dispute and claims, from which the greatest benefit went to the lawyers.

6.1.2 Global trends

Present utility organisations, many now privatised, tend to have greater independence. Those that remain in the public sector are being made more accountable. Bidding is becoming more competitive. Frequently commercial organisations carry out long-term management contracts for utility operations. In all cases, greater pressure is put on utility organisations to provide a higher standard at a lower cost.

Utility clients are therefore questioning traditional contracting arrangements and the parts played in them by the engineer and contractor. Some clients have commercial interests in engineering and construction groups. Generally they are looking for integrated solutions and guarantees on time, price and performance. The relationship of engineer and contractor may be one of combining their functions as partners in a project rather than to work separately.

6.1.3 The changing nature of work

In industrialised countries, the demand has shifted from extending existing pipe and cable networks to rehabilitating old ones, often long neglected or at least past their design life. Almost by definition these are in the older, more congested parts of town. Earlier assumptions of the engineer's design and the contractor's approach that an open trench method would be the most convenient and economic are less applicable in city centres than they were in greenfield sites. Open-cut trenches are never convenient in urban areas and, often uneconomic even on a direct cost basis. When the indirect costs of environmental damage and the social costs to community well-being and commerce are added, trenching can be very uneconomic indeed.

Clients are more aware, not only of the disadvantages of traditional trenching methods, but also of the possibilities offered by trenchless technology as an alternative. The impact of public (and often, shareholder) opinion, legal restrictions on working and available information on comparative costs also have an increasing influence on their decision-making and attitude to design.

6.1.4 Integrated design

For pipe installation, the design and method chosen for implementing the work are inextricably linked. It is unrealistic to believe it is possible to plan and engineer works without taking their construction into account. Sidestepping the issue by allowing alternative bids is really not an answer. One form of construction cannot be substituted for another at bid stage without compromising the technical and economic gains that could have been achieved by an integrated design. A growing number of clients now appreciate that, in order to exploit fully the advantages of trenchless installation, the concept must be introduced early into the design process.

Pipejacking and microtunnelling provide the designer with a degree of freedom not available with open cut. For example, surface features such as street furniture or trees do not have to be avoided. Unlike open cut, increased depth of installation adds only a small percentage to costs. This can be a major consideration in designing a scheme.

However, more is required of the designer. He must be able to evaluate the relative technical merits of available solutions and assess their relative costs. This means knowing a great deal more about trenchless options like pipejacking and microtunnelling than he will learn from discussions with equipment vendors or contractors' representatives.

This concept of linking design and method of construction runs counter to many deep-seated ideas. Many fear that to design a sewer scheme using microtunnelling techniques would expose engineers to additional professional liability. Some clients feel they will be criticised for restricting the number of contractors able to bid. Albeit subconscious, a traditional view is that a low-technology solution, trenching, allows the maximum number of contractors, even the inexperienced, to bid. The engineer can plead precedent if anything goes wrong.

Design of trenchless solutions requires more planning and preliminary preparation Recognising this, some clients, engineers and contractors have turned to different forms of contracting.[See Chapter 12]

6.2 Conceptual design for sewer installation

Certain criteria have to be considered at the initial stages of planning any pipeline. These include:

- Hydraulic requirements
- Corrosion and durability
- Route alignment
- Access restrictions and requirements
- Line and grade
- Design depths
- Maintenance and repair
- Geological and hydrogeological considerations
- Community and environmental impacts
- Hazards and risks

The installation of gravity sewers, the primary application for pipejacking and microtunnelling, must allow for some additional criteria because of the specific characteristics of sewers themselves. This section is concerned with the way in which the choice of pipejacking or microtunnelling can affect planning and design for a sewer installation.

Although the overall route location and hydraulic design of pipes lie outside the scope of this book, the design of jacked pipes, including corrosion and durability issues, are discussed in Chapter 10.

6.2.1 Line Installation: alignment, gradient and depth

Sewers are laid to serve designated areas and the properties within them. Access to the sewer pipe has also to be provided at appropriate and convenient intervals for the purposes of future cleaning, maintenance and repair. Sewers are laid at greater depths than other services and at gradients which allow for gravity flow. Hydraulic considerations therefore require that pipes be laid closely to specified invert levels and as far as possible to straight lines. To meet these requirements and to avoid other utilities, sewers are most often located in the road, with access shafts at junctions and at changes of direction.

The design of the alignment must account for rights of way, easements for going under property and surface features as well as existing utilities and other underground structures. The choice of route and space available may be severely limited where the object is to replace an existing system in a developed residential area. Site visits by a competent engineer are essential at the preliminary stage. A trained eye will note significant physical, environmental and man-made features. In suburban areas it is likely that many surface features do not show on records and drawings. Current excavations on local construction sites are a useful source of information.

Jacked methods and the influence on alignment. With jacked methods more alignments are feasible than with open cut [35]. On numerous occasions jacked solutions have reduced the installed length that would have been needed for an open-cut route. Whereas an open-cut alignment may have been diverted away from a street with mature trees to meet a local authority requirement, a shorter direct route could be allowed by jacking underground. Jacking allows both vertical and horizontal curves to be driven, although straight alignments are easier to install and would be preferred for the hydraulic design. An extreme example of a jacked curved sewer to keep it within the boundaries of a Japanese street is reflected in the sewer alignment shown in Figure 6.1.

The safe radius that can be installed by jacking depends on a number of factors including the section length of pipe, type of joint and ground conditions. Where no special measures are taken, it has been suggested that the safe minimum radius should be restricted to 300 times the diameter. With short-length pipes and joints capable of accepting greater angular deflection, the multiplier can be reduced.

Figure 6.1 Curved jacking (photograph by courtesy of Showa Civil Engineering, Tokyo)

Shaft location. Wherever possible, drive and reception shafts are positioned in the same place as the permanent access shafts. Access manholes for sewers are spaced according to operational and maintenance requirements. A commonly used rule is for access every 100 m for non-man-entry sewers, but authority requirements vary. Access points for man-entry sewers can be spaced further apart. As practical drive lengths for jacking methods increase in relation to pipe diameter, few situations occur where the required shaft spacing cannot be met. In practice, spacing for smaller pipejacking diameters is determined by economy or safety rather than by technical limits.

Drive lengths of more than 1000 m have been jacked for diameters of 1800 mm or greater. For pipes in the 1500 mm diameter range, drives of 500 m are possible, but for safety reasons spans need to reduce as pipe diameters decrease. For 1200 mm diameters, around 250 m is a sensible practical limit.

When positioning drive shafts, the designer needs to take construction activities into account. Shafts must be placed to allow room for their construction and for the temporary top side equipment to be set out, although fortunately this is usually flexible enough to be arranged to fit particular site limitations. Access must be provided for trucks to bring in pipe sections and take away spoil. Where possible drive pits are used for driving in two or more directions. Disturbance to the environment and conflict with the public and traffic also need to be considered when shafts are planned. The benefit of trenchless pipelaying is lost if the temporary occupation and shaft works create as much obstruction and disruption as traditional working.

The current trend is to use standard precast circular shafts for drive and reception which can subsequently be adapted to serve as the permanent access manholes. This integrated approach brings many benefits, and is a further example of how permanent and temporary design can be co-ordinated. Chapter 11 discusses shaft design in more detail.

Installation tolerance. The greatly improved capability of modern jacking technology to install pipes to close tolerances, typically within ±25 mm, compares favourably with what can be practically achieved in open cut. The latest improvements enable tolerances to be reduced still further, in some cases to ±15 mm.

Depth of installation. Depth of cover to the pipeline is both a technical and an economic issue. Chapter 14 reviews the economic considerations; Chapter 8 considers the influence of loading on pipe design.

For sewer design, one clear advantage of underground methods over trenching is the low cost increase for deeper installation. Whereas in open-cut, working costs increase almost geometrically with greater depth, with jacking the only additional cost is for the deeper shaft. With a deeper installation the number of pumping stations can be reduced, which can save on capital and running costs of the system. Alignments can be taken under areas of high ground which would be prohibitively expensive by open cut.

Other practical depth issues that the engineer must consider are the minimum operating depths for microtunnelling and pipejacking operations. Pressure balance methods require a minimum depth of cover to hold pressure. A figure often suggested is 2 m to pipe crown, but this may need to be greater in open granular materials or where there are fissures. Loss of pressure to the surface can cause face collapse and also surface pollution by slurry. Open-shield systems, including auger microtunnelling, require less depth of cover, drives with as little as 1 m of cover have been completed without surface disturbance.

6.2.2 *Subsurface considerations*

When alternatives are being evaluated at the design stage, it is essential to know what is likely to be lying below the surface. Sound information will greatly help determine the nature and extent of the ground investigation programme. A great deal can be learned about ground conditions from existing sources of information on local geological and geotechnical data and from historical records.

Geological data. Geological records and maps reveal subsurface materials and how they were deposited, transported or created. Soil information will show whether it is organic, residual, aeolian, colluvial, flood plain, alluvial, lacustrine, or glacial moraines. Other characteristics like the incidence of boulders, overconsolidated clays or cemented layers will also be significant, as will the type of bedrock and the depth to bedrock. Hydrological data on ground water regimes and the location of any artesian or perched water conditions will be essential. Geological features can then be identified on site visits and the information interpreted.

Geotechnical data. Most town or regional authorities maintain a data base of borings and soil investigations undertaken in the past. Local ground investigation firms may assist by allowing access to their records. Reports of previous local activities provide more detailed information on soils and water tables encountered and changes in conditions.

Historical records. Local records can show how certain areas have been used in the past; the impact of man on natural conditions. Important features might include diverted water courses, old quarries or ponds subsequently filled. Sites of former factories, works, depots or tips in the area may indicate the presence of contaminated materials underground. By the same token, the existence of major utilities and other underground constructions, like subways, basements, chambers and foundations, will influence planning and design.

6.2.3 *Construction considerations*

Geotechnical factors and construction method. The presence of a high water table is a problem in many soils. For traditional trenching two options are available for dealing with this. One alternative is to stabilise the ground by driving interlocking steel sheeting. This can be extremely expensive and damaging. The other is to lower the water table by pumping to allow trenching work to proceed. This is possible in granular soils but can remove fine particles from the soil and cause settlement of property and other pipelines. Lowering the water table can also have harmful effects on trees and shrubs. Some cities are therefore restricting or even banning techniques which require the ground water to be lowered.

Pressure balance types of pipejacking and microtunnelling have the advantage of being able to work in unstable soils and high water tables. However, where the designer has a choice of soil conditions at different depths, it is preferable to jack in consistent rather than mixed ground. It is more economic to drive through consistent water-bearing silt and sand than through clay soil with intrusions of silt and sand.

Where polluted soils are suspected, remote-control closed pressure systems should be considered as these can offer several advantages:

- The material is isolated from human contact.
- The volume of polluted material to be disposed of is minimal.
- The need to extract and treat polluted ground water is eliminated.

6.2.4 Traffic, social and environmental considerations

The potential impact of works on the community and the environment is another factor which must be considered at the design stage. With increasing demands on the space above and below the highway, the rights of other users have to be taken into account. The greatest conflict occurs when construction occupies part or all of a heavily trafficked road. Until recently it has been possible for the owner and his designers to ignore the costs of traffic delay and commercial disruption and to aim for the method giving the lowest site cost. In many countries new legislation or restrictions now limit the utilities' freedom of operation in the road. Strict requirements in a city like Tokyo [36] have made open cut trenching increasingly unattractive and expensive, encouraging the use of trenchless methods.

Various schemes now ensure that owners and their designers account for the true costs of a sewer installation. Road rental and charging for road space have been introduced in a number of countries. Although a provision in the British 1991 New Street Works Bill [37] to charge for the occupation of road space during utility works has not been implemented, reserve powers are still held. Whether this clause remains dormant or not will depend how the utilities cooperate in reducing the impact of their works..

Until recently the impact of construction works on the environment was not an issue for most utilities and their designers. This has changed dramatically. They now have to spend considerable time and effort finding ways of re-routing or going under to avoid disturbing or damaging sensitive areas.

Sewer construction is undoubtedly the most disruptive form of utility installation. Studies [38] have shown that the cost of traffic delay for a sewer installation in an urban road can be two or three times the value of the works. In one case the estimated social costs arising from a difficult open-trenched sewer rose to 10 times the value of the work.

The chosen method of construction is going to be the main determinant of the degree of impact a scheme has on traffic, community and environment. The engineer will have to produce a fully-integrated design solution, which includes details of construction, if he is to satisfy not only the clients but also the community. Evidently less disruptive, pipejacking and microtunnelling will satisfy these demands so long as the job is designed and laid out in a way which accommodates traffic and inconveniences the neighbourhood as little as possible. Nor do these trenchless methods only offer advantages during the installation; problems associated with long term settlement and pavement reinstatement can be minimised and environmentally sensitive areas can be passed under. In addition, damage to existing underground utilities is less likely, because the greater part of the work will be at depths below most pipes and cables. The zone of influence and disturbance of a jacked pipe is also less than one laid by trenching.

6.3 Location of crossings

6.3.1 Planning and design

The planning and design of a crossing involves the same design considerations as for line installation. When attempting even the most elementary analysis it becomes obvious that any form of failure is going to create unacceptable risks and costs. Considerable thought must therefore be given to the siting of crossings and the likely ground and hydrological conditions. Installation methods selected must be able to cope with the widest range of conditions likely to be met. As crossings are made at critical locations under roads, railways, waterways or buildings, failure or even partial failure, such as loss of soil, can be dangerous and expensive to remedy. There is usually little opportunity to sink exploratory or recovery shafts for remedial measures.

Any additional limitations imposed by the owner of the structure being crossed have also to be taken into account. These vary. Some regulations imposed by rail and highway authorities, for example, were formulated some time ago. Requirements do not therefore reflect the state-of-the-art performance of pipes at crossings or the technology available for their installation. Some authorities continue to call for a steel casing at a crossing, a requirement now considered redundant by many researchers. Valuable work undertaken by Cornell University and sponsored by the American Gas Research Institute (GRI) addressed the loads and design of pipes at crossings. The results of this research, reported in several papers at a Conference in Denver 1991 [39], showed that casings were often unnecessary.

6.3.2 Investigation

Site investigation of the soil conditions has to be limited to adjacent areas rather than to the crossing itself, which can provide a false picture of what actually lies below the road or railway. In France a pipejacked crossing had to be aborted when the remains of a locomotive were encountered. It transpired that during the Second World War the rail track had suffered a direct hit from a bomb and a locomotive had run into the crater. The occupation troops had decided it was not worth recovering and had backfilled around it.

Foundations, like basements and piles, must be investigated thoroughly if a crossing is to pass under a building. One embarrassed authority had to abandon a TBM and a section of tunnel when the machine ran into a deep piled foundation. A second lesson can be learned from this incident: the building was recent and constructed after their line survey which had not been updated before work started.

6.3.3 Crossing waterways

Difficult ground conditions and high risks can attend crossings under waterways, rivers and canals. Northern Europe is networked by navigable waterways and drainage canals. To carry a pipeline under them may mean driving several hundred metres at considerable depths to ensure a safe cover under the bed. Work is likely to be in difficult ground with a high head of water.

Several companies have developed jacking technology to meet these conditions. Drives with a vertical curve which forms an arc profile can reduce shaft depth, and the associated costs. See Figure 6.2. Some form of pressure balance shield of the remote or compressed-air type is required.

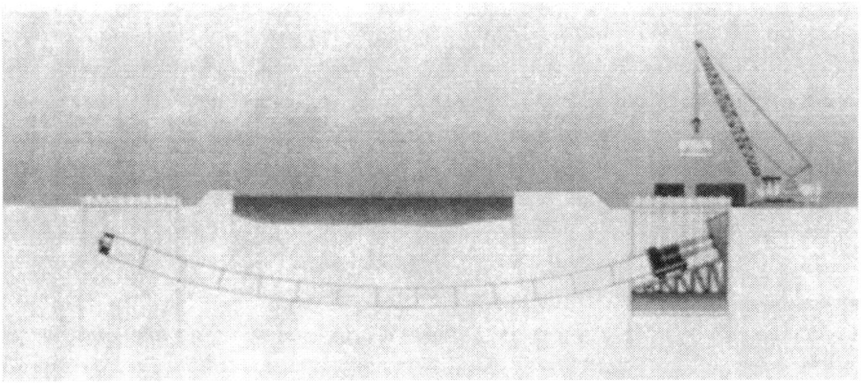

Figure 6.2 Arc river crossing (source: Denys, Belgium)

7 Subsurface and site investigation

Both subsurface and site investigations are prerequisites for the design of any underground construction. The results of these will confirm (or otherwise) geological data and other underground information collected at the planning stage and provide the data necessary for the final design.

7.1 Subsurface investigation

Apart from confirming the findings of the site investigation, the subsurface investigation quantifies soil properties and ground water regimes. This information, fundamental to the safe and efficient design of pipejacking and microtunnelling works, will be the basis on which decisions are made about the following elements of the design:

- Behaviour of subsurface and choice of shield type
- Choice of cutting heads
- Spoil removal method
- Fluid system - water/mud/polymers
- Calculation of jacking loads and number of IJSs
- Lubrication and jacking load reduction
- Methods of shaft construction
- Spacings for shafts
- Shaft jacking capacity
- Shaft dewatering
- Reaction walls

Methods of obtaining detailed information about soil and hydrological conditions include:

- Drilling, bore hole sampling and laboratory testing of recovered samples
- Field testing
- Non-destructive testing
- Geophysical techniques

In most countries, guidance is provided in codes of practice and standards for testing, though the methods of testing are much more detailed than the extent of the recommended investigations.

7.1.1 Cost of investigation

The US National Committee for Tunnelling Technology studied the procedures for site investigation and their influence on underground work [40]. From the evidence came two main observations. First, the projects which had the greatest length of investigatory boreholes had the final contract cost closer to the bid price than those with less investigation. Second, those with the better investigation had the fewer claims. Among other things, the Committee made these recommendations:

(i) The level of site investigation cost should be about 3% of the estimated construction cost. This compares with the figure of 1% frequently quoted by engineers and clients.
(ii) The amount of investigatory bore-holing should be 1.5 times the tunnel length.

Although the study was concerned with tunnelling, the comments appear to be equally valid for pipejacking and microtunnelling.

7.1.2 Borehole sampling and laboratory testing

Bore-holing and the removal of disturbed and undisturbed samples, followed by a series of prescribed laboratory tests is the classic form of investigation.

Boreholes. Boreholes should be located at all shaft points and intermediate borings at every 80-100 m. If highly changeable conditions are found, it will be necessary to make additional boreholes and/or use other means of investigation and to correlate the results.

The correct siting of a borehole is just off the line. The incorrect siting and refilling of boreholes is a common fault, as many tunnellers and jacking contractors have found to their cost. As a rule, boreholes sited along the line need to be taken down below the invert by about 2 m or by 2 diameters, whichever is the greater. Continuous sampling from one diameter above and one diameter below the proposed pipe is recommended.

Borings at shaft locations, however, should be within the proposed shaft area. Such borings should be taken down about 4 m below the proposed subgrade level. For shafts of diameters 5 m or greater, this depth will increase with diameter.

The physical difficulties of site investigation in congested areas can be substantial and include:

- Obtaining permission to bore and investigate.
- Avoiding conflict with traffic.
- Adhering to safety precautions.
- Investigating in a congested network of services.
- Avoiding buried structures and problems with disturbed ground.

Locating bores for crossings presents problems. Bores should be taken at drive and reception pit locations, and ideally also in between. Obtaining permission for such a borehole is unlikely to be forthcoming.

Investigations must be conducted with particular care for waterway crossings, particularly beneath rivers. Rivers are living entities; the beds will have changed over a period of time, so information obtained from boreholes on the banks may bear no resemblance to what is found below the bed. Useful information on investigation for river crossings will be found in Charles Hair's papers [41] [42].

Laboratory testing. Laboratory tests on disturbed and undisturbed samples would generate the following categories of information:

(i) Physical properties
- a. Mineral identification
- b. Density
- c. Density limits
- d. Water content
- e. Gradation characteristics
- f. Particle shape characteristics
- g. Consistency limits

(ii) Hydraulic properties
- a. Permeability
- b. Piping potential
- c. Characteristics of pore fluid

(iii) Mechanical Properties
- a. Compaction limits
- b. Consolidation properties
- c. Elastic constants
- d. Shear strength (various forms)

From this mass of information an experienced geotechnical engineer can identify and describe soil type, and classify and quantify its properties. However, it should be noted that the results obtained are those for the samples taken, not for the actual soil in place. The securing of the samples, their encapsulation, handling and reconstitution may create differences between the sample and the soil in situ. It should also be noted that some of the tests in reality only provide a comparison between one sample and another, rather than a picture of the likely soil behaviour. Unfortunately, soil classification systems can vary from one country to another, or even between different authorities in the same country. Soil descriptions for the same material can vary from one system to another. The *Unified* system, or variations of it, is the most common.

The interpretation of the results needs particular skills and experience. Information obtained from other sources also needs to be taken into account and, where significant inconsistencies occur, careful cross checks should be made.

Drilling and sampling have acknowledged shortcomings, which include the following:

- Laboratory testing of bore hole samples is slow to yield information. A bore hole may take several days to complete; more days may elapse before the bore hole log is available; it may take weeks to get the test results from the laboratory.
- Samples from drill tools are often of poor quality. Strata logs can be misleading because larger gravel and cobbles will not enter the sampler..
- Drive sampling in some types of ground can yield inadequate results.
- Rotary coring in rock, for example in fractured strata, can be of poor quality because the use of wash water for cooling the bit greatly alters the specimen.

Methods which overcome some of these deficiencies include rotary drilling and boring in reverse circulation.

Rotary drilling. Large diameter wireline drilling up to 150 mm appears a successful method of coring over-consolidated cohesive ground and weak rocks. Triple-tube barrels are used and the bore is fully lined. Cores are retained in a removable plastic inner liner Using polymer fluids with wireline drilling has helped core recovery in cohesive and cohesionless soils. The polymer fluid adds support to the bore hole wall and assists logging by forming a thin skin at any discontinuities.

Reverse circulation boring. A flushing medium is pumped through the annulus between inner and outer concentric tubes. Samples are collected via the inner tube, which is fitted with a non-return valve. Improvements include a drill bit which can penetrate and take samples from unconsolidated strata and a *non-return* valve, a ribbed and petalled plastic disc, which restricts intermixing and loss. Reverse circulation boring gives samples of better quality. However, it is not suited to all types of soil and the equipment requires greater operator skill.

In many cities, conventional geological information may be of little use unless construction is quite deep. There is often a substantial depth of made or disturbed ground including rubble and old foundations. In the centre of Milan, for example, the depth of this is 3 m, and often considerably more.

7.1.3 Borehole and field testing

Various field-testing methods can provide valuable information, especially when cross-referenced to other sources. When boreholes are drilled, a number of field tests are made, including visual observation of samples and tests of strength and ground water levels. Permeability can be determined by observing either constant or falling head type tests. The Pressuremeter Test, though more complex, can yield considerable information about deformation modulus and other soil characteristics.

Many field tests provide data on soil strength. Tests range from using thumb and pocket knife, through hand vanes and penetrometers, to sophisticated mechanical equipment. These include the universally known Standard Penetration Test from which N values are obtained. (N values are the number of blows required to drive a standard sampler 1 foot, using a standard size hammer falling a standard distance). A number of variables can influence this result. Blow count has been roughly correlated to many properties such as modulus of elasticity, cohesion in fine-grained soils and friction angle in granular soils. Table 7.1 sets out a scale of N values and their relation to strength and soil description. Practical rule-of-thumb site tests and the relationship of the information obtained are also shown.

Table 7.1 Scale of strengths and N values for weak rocks and soils (after Robb, *Site Investigation*, Thomas Telford, 1982.)

Unconfined compressive strength: MN/m²	Shear strength: kN/m²	Approx. N value	Strength / consistency	Grade	Breakability	Penetration	Scratch
100	40 000						
			Strong	A	Difficult to break against solid object with hammer		Cannot be scratched with knife
50	20 000						
	10 000	600	Moderately strong	B	Broken against solid object with hammer		Can just be scratched with knife
	8000						
12·5	6000						
		400					Scratched with knife. Can just be scratched with thumb-nail
	4000			C	Broken in hand by hitting with hammer		
			Moderately weak				
5							
	2000	200		D	Broken by leaning on sample with hammer	No penetration with knife	Scratched with thumb-nail
			Weak				
	1000						
	800			E	Broken by hand	Penetration to about 2 mm with knife	
1·25	600	100					
	400	80 60	Hard or very weak	F	Easily broken by hand	Penetration to about 5 mm with knife	
	200	40	Very stiff			Penetrated by thumb-nail and to about 15 mm with knife	
			Stiff			Indented by thumb	
	100	20					
	80						
	60		Firm			Penetrated by thumb with effort	
	40	10 8					
		6	Soft			Easily penetrated by thumb	
	20	4					
	10	2	Very soft				

7.1.4 In-situ test methods

Among the most widely used in-situ testing techniques are the static, or Dutch cone, which allows determination of soil type, strength and degree of compactness and the piezocone, which allows measurement of pore water pressures.

The static cone. The static, or Dutch cone, is used to carry out the cone penetration test (CPT). In the CPT, a cone on the end of a series of rods is pushed into the ground at a constant rate, and continuous or intermittent measurements are made of resistance to penetration of the cone. If required, measurements are also made of either the combined resistance to penetration of the cone and outer surface of the rods or the resistance of a surface sleeve. The CPT is used in Europe, the USA and elsewhere.

The CPT has three main applications:

- To determine the soil profile and identify the soils present.
- To interpolate ground conditions between control boreholes.
- To evaluate the engineering parameters of the soils and to assess bearing capacity and settlement.

In relation to certain problems, the third application, the evaluation, is essentially preliminary in nature, to be supplemented by borings and by other tests, either in-situ or in the laboratory. The CPT, in this respect, provides guidance on the nature of such additional testing, and helps to determine the positions and levels at which in-situ tests or sampling should be undertaken. Where the geology is fairly uniform and predictions based on CPT results have been extensively correlated with performance (as in parts of Belgium and the Netherlands), the CPT can be used in ground investigation. Even in these circumstances, CPTs may be accompanied by, or followed by, borings for one or more of the following reasons:

- To assist where there is difficulty in interpreting the CPT results.
- To further investigate layers with extremely high or low cone resistance.
- To explore below the maximum depth attainable by CPT.
- To be used if the project involves excavation to obtain samples required for laboratory testing and where knowledge of ground water levels and permeability is needed.

Depth of penetration is controlled by the ground conditions. Light rigs penetrate clays and silts of firm consistency, and go some way into loose and medium dense sands. Heavy rigs penetrate stiff clays and loose gravels, but make little headway into dense gravels. Cobbles and boulders can prevent further progress, and progress in rock is limited to the weathered upper surface.

Soils are identified by making empirical correlations between soil type and the ratio of local side friction to cone resistance (skin friction ratio) considered in relation to the cone resistance. Engineering parameters are also assessed on the basis of empirical correlations. Alternatively, the test results can be directly used to estimate bearing capacity and settlements, again on an empirical basis. The CPT has advantages over the usual combination of boring, sampling and standard penetration testing (SPT):

- It provides a continuous, or virtually continuous, albeit indirect, record of ground conditions.
- It avoids the ground disturbance associated with boring and sampling, particularly that which occurs with the Standard Penetration Test (SPT).
- Any disturbance resulting from the advancement of the cone is consistent between one test and another.
- Cost savings can be made by using the CPT.

In practice the use of the CPT remains essentially empirical. The main relevant parameters are angle of shearing resistance and deformation characteristics in cohesionless soils, and undrained shear strength and modulus in cohesive soils. One of the important uses of the CPT is to delineate the soil profile. This it can do with greater accuracy than can be achieved from conventional boring and sampling.

The Piezocone. Piezocones have been introduced which combine pore water pressure measurement with the measurements of both cone resistance and skin friction. Applications of the piezocone fall into two groups — for profiling and for assessing engineering parameters.

Profiling with the piezocone gives more accuracy than the CPT, but as yet it has not led to any major improvement in identification of soils. The main practical application in assessing engineering parameters is for measuring static pore water pressures.

Piezocone sounding is particularly useful in mixed deposits where it is often difficult with the CPT to know whether the data refer to drained or undrained conditions.

A number of other cone penetrometer devices have been developed.

7.1.5 Geophysical techniques

Geophysical methods of investigation fall under the six main heads of electrical, seismic, electromagnetic, radar, magnetic and gravimetric surveys. They can be used to help interpret geological and hydrological conditions along the proposed alignment In practice, seismic methods are the most developed and most widely used method for assisting in the evaluation of ground conditions. But radar is the system on which the greatest amount of research effort has been expended. Many of the methods are used for locating underground objects, including pipelines and cables. These are discussed in the next section.

Seismic. Seismic methods exploit the differing characteristics of soils and rocks that affect the speed with which vibrations are propagated through them. They can work with a range of vibration sources including resonance, reflection, refraction and high frequency. Refraction methods are used for shallow exploration in the top 100 m of the surface.

Seismic investigation can be done by surface or borehole techniques, singly or in combination. The waves can be generated by a small explosive charge or by hitting a metal plate with a hammer. Geophones are positioned at the surface roughly at intervals corresponding to the depth of investigation required. The geophones detect the refracted waves and transmit the signal to a seismogram, which produces a trace of movement and time.

Seismic methods can be valuable for identifying major changes in strata. Profiling bed rock or sub-surface features, like caverns or large boulders, are typical applications. It is important that the results are closely correlated to bore hole investigation and, if necessary, checked with probes or additional boreholes in critical situations. Experience shows that the strata profiles indicated by seismic waves do not always correspond with borehole information.

A full explanation of the theory and techniques of seismic wave investigation can be found in *Geology for Civil Engineers*, by McClean and Gribble [43]. A seismic system being developed by NTT for surveying relatively shallow routes to provide strata information is described by Kuroiwa [44].

Ground probing radar. Ground Probing Radar (also known as Ground Penetrating Radar and Ground Scanning Radar) involves transmitting into the ground a pulse of electromagnetic radiation via a directional antenna, and measuring the reflected return signals to indicate the depth and location of objects relative to it. Signals are then transmitted to a control unit where they are amplified and filtered before digital processing. However, ground probing radar (GPR) has some practical limitations:

- The system is complex and difficult to operate.
- The depth of penetration is limited (about 5 m) and object resolution decreases with depth.
- The interpretation of data requires experts.
- Varying soils will have differing relative dielectric constants of the ground used by the computer software to calculate depth. These constants will have to be established in the area being surveyed.
- Penetration is limited in wet soils as water is a great absorber of this type of electromagnetic radiation.

The broad consensus of opinion is that at this stage of development the quality of information that can be obtained is not sufficiently reliable. Berosch reports some success with GPR for detecting geological anomalies on a site in Paris [45].

Other methods. Other sub-surface investigation methods have yet to find any mainstream application, but a description of their potential use and development work on electric and gravimetric methods can be found in a paper to the Paris No-Dig Conference, 1992 [46].

7.1.6 Information from testing

The three most valuable items of information from testing concern:

- The strength for clays - undrained shear strength and density for granular soils.
- The location of ground water table, particularly in granular soils.
- The variation of soil types along the line.

Discussed below are some points worth noting about different soil types.

Cohesive soils. Short-term strength, the *undrained shear strength*, is determined in the laboratory from undisturbed samples of sufficient size. In heavily overconsolidated or fissured clays, small samples can show an overestimated undrained strength. Readings from cone penetration tests can be converted to strengths. However, results for clays should be translated carefully as many of them are anistropic, that is the strength measured in the horizontal axis is different from that measured on the vertical axis. Undrained shear strength can be measured in trial pits by using penetrometer, torvane or shear vanes.

Atterberg limits (plastic and liquid limits) and natural moisture content can only be determined by test. If the clay content (percentage finer than 2 microns) of highly plastic clays is determined, and this information is combined with Atterberg values, an initial assessment can be made of potential heave and swell.

Granular soils. Particle size distribution (PSD) and the chemical characteristics of soil can be determined in the laboratory. These characteristics, along with measurements of site permeability, are likely to provide the only useful information for granular soils. In PSD, the proportion of fine material is important because permeability is determined by the finer 10% to 20% of the soil particles. Any cobbles or boulders encountered will not be retrieved through borehole techniques. They should be recorded on a driller's log and be taken into account in assessing the maximum particle size. Provided the PSD of the soil is determined over the full range of particle sizes, the coefficient of permeability can be related empirically to PSD. Where piezometers have been installed, the permeability of fine-grained soils around the piezometer tip can be determined by using rising or falling head techniques.

Rock. For harder rock (unweathered or low weathering), samples are retrieved by rotary drilling with an air- or water-flush core barrel. Fracture spacing or RQD (Rock Quality Designation) can also be logged. RQD is the percentage ratio between length of recovered core pieces greater than 100 mm and the total core drilled. Table 7.2 sets out the descriptive terms used for RQD. Its compressive strength is determined by testing. Descriptive terms for compressive strength are listed in Table 7.3. For softer rocks, such as marl or chalk, the weathering grade is of importance and trial pits are more valuable than bores.

Table 7.2 Descriptive terms for RQD values

RQD (%)	Description
0 - 25	Very poor
25 - 50	Poor
50 - 75	Fair
75 - 90	Good
90 -100	Excellent

Table 7.3 Comparative strength of rock material

Compressive strength (MN/m^2)	Description
<1.25	Very weak
1.25 - 5.0	Weak
5.0 - 12.5	Moderately weak
12.5 - 50.0	Moderately strong
50.0 - 100.0	Strong
100.0 - 200.0	Very strong
>200.0	Extremely strong

7.2 Site investigation

7.2.1 Types of investigation

Utility authorities, engineers and contractors need to know the siting and state of the existing underground network and also of abandoned pipes and cables. They also need details of other manmade underground structures, like foundations, underground chambers, buried ruins, voids or even unexploded bombs. Information can be obtained by researching existing records, by inspecting the site and by various means of site investigation.

Records. Local records are only reliable if they are accurate and up to date. In urban areas particularly, utility installation has been going on for nearly 200 years. Not surprisingly, records for some were either never made in the first place or have been lost or destroyed. It would also be optimistic to assume that records have been meticulously updated whenever changes have been made because of damage, maintenance or diversion. Nevertheless, good local records can be a valuable information source.

Site inspection. Surface signs, such as covers and chambers, markers and traces of reinstatement will indicate the line of the various utilities. However, pressure pipe or cable may diverge appreciably from a straight line between two fixed points. Assessing the depth of services can be more difficult, but generally standard requirements for depth of cover will have been followed. Authorised lifting of covers can provide useful depth information, particularly on sewers.

Site investigation. Trial-holing is the most basic way of locating existing pipes and cables. More sophisticated location devices can give much more accurate readings. These are based on one of four main principles: electromagnetism, acoustics, radar or infrared thermography. They can be used at the survey stage where the objective is to trace and locate, and during construction where the objective will be to avoid.

Elecromagnetic devices are the most commonly used for locating utilities [47]. Metallic pipe and cables can be relatively easily traced. For non-metallic pipes, a sonde, inserted into the line, can be tracked by surface locators, a method particularly useful for revealing the line and depth of sewers.

One use of acoustic energy has been in the application of the water-hammer phenomenon; by inducing an energy impulse into plastic water pipe, the signal can be picked up by a seismic sensor.

Ground penetrating radar (GPR), used for soil investigation, has been developed for locating existing utilities [48]. Despite considerable progress with this method, some reports suggest that a success rate of only about 90% is achieved.

Infrared thermography has not been pursued for utility location, as it is found to be over-sensitive to external conditions.

7.2.2 Points of conflict with underground structures

Pipejacking and microtunnelling are most likely to conflict with buried utilities and other underground structures in three ways:

- By interfering with buried structures in the area of shaft construction.
- By directly encountering other structures during installation.
- By indirectly damaging structures through ground movement they may cause.

Around shafts. Defining and locating utilities in and around shafts is relatively straightforward, requiring normal search techniques, but trial pits may be needed for confirmation. Quite frequently existing pipes and cables have to be diverted out of the area of a shaft.

Direct encounter. More serious is running directly into an existing pipe or structure. Apart from the consequences of any damage to the structure itself is the problem of how to deal with an immovable object directly on the alignment. Even if it is possible, diverting the existing pipeline will be time-consuming and expensive. The alternative, also costly, is to abort the job, which may mean losing the TBM, and to find another alignment for the new pipe. Prevention, through sound investigation and planning, will be better than cure. Specialist equipment and operators may be needed to search depths greater than 5 m. As location becomes more difficult with increasing depth, more reliance has to be placed on utility records.

Fortunately, few utility lines are laid at sewer depths. Investigation can therefore be concentrated on those pipelines that may be at greater depth. These are likely to be storm and gravity sewers and transmission mains for gas and water. Fibre optic cables are also being placed deeper to provide greater protection. As these are valuable assets and of recent origin, records are usually good. However, over the years, a number of deeper strategic pipes and cables have been laid by governments and military forces. Records of their whereabouts may not be in the public domain, but municipal authorities usually have a good idea of their existence if not their precise location.

Indirect damage through ground movement. The need here is to ascertain the condition of existing pipes. Any movement or disturbance can cause a corroded or rigid pipeline, such as cast iron, to fail or leak. Information on condition is not always known or readily available from the pipe owner. However, knowing the age, type of pipe and the soil conditions will be a good start. Experience will be a blessing.

Well-chosen and skilfully-operated pipejacking and microtunnelling systems have advantages over trenching or even tunnelling methods in minimising ground movements. For this reason jacking methods have been specifically selected to cross in close proximity to foundations or existing pipelines. Chapter 9 discusses the choice of shields to suit particular ground conditions. However, displacement methods should not be used close to existing structures as they create movement in the surrounding soil. How close will depend on the soil conditions and the amount of soil being displaced, but a suggested minimum is for a clearance of 1 m.

8 Design: permanent works

8.1 Pipes and pipe connections

8.1.1 Requirements for jacking pipes

The pipe and the access shafts are the only permanent parts of a pipeline construction and the parts the client is paying for. The cost of the pipes and shafts, particularly for a deep sewer construction, represents only a small percentage of the total cost of the installation. The larger percentage by far is the cost of the temporary works: the mobilisation of equipment, the excavation and restoration. This high cost of placing the pipe to the required position in the ground, together with the appalling consequences of failure, should be reason enough for designers to select materials of the highest quality which meet the technical requirements for their purpose. Engineers who are swayed by the initial cost of materials rather than their long term value do little but contribute to the growing volume of sewer rehabilitation of the future.

The selection of pipes and their connections which will offer cost-efficient sound performance throughout their design life (30-100 years) therefore demands a great deal of technical knowledge on the part of the engineer. Fundamental requirements of pipes are:

- Resistance to internal and external corrosion.
- Capacity to withstand static and dynamic loadings.
- Capacity to withstand internal and external pressures.
- Good flow characteristics.
- Satisfactory whole-life costs.

Pipes used for pipejacking and microtunnelling are thick-walled to withstand the jacking loads and are precisely fashioned to close tolerances. They are high strength, high quality products which will withstand the live and dead service loadings for most situations. In addition to satisfying the normal requirements, their specific design features include:

- High axial load capacity.
- Close dimension tolerances.
- Squareness of ends.

- Straightness along the length.
- A watertight joint made within the pipe wall, without any internal or external projections.
- Joints capable of transmitting axial loads while remaining watertight under angular deflection.

Some recent microtunnelling techniques do not apply jacking loads to the pipe end, but distribute the loads along the body of the pipe. These improvements have made it possible to install pipes with a relatively thin wall for microtunnelled sewers. However, with thinner pipes the service loading becomes more critical. National standards for various types of jacked pipe now exist in many countries.

Section length. Typically a standard pipe is 2.0 to 3.0 m long, with short sections in 1 to 1.25 m lengths, although larger diameters are sometimes in greater lengths. The length of section used will be determined by the relative advantages of the reduced number of joints to be made for standard length pipes against the advantages of reduced dimensions for the working shaft required for short length pipes.

It has been suggested that the length of individual pipe sections should not be greater than the length of the shield. It is argued that, as a jacked installation advances by a series of small deviations and corrections, pipe sections of a greater length will not be able to follow the course of the shield. Contact with the soil will be increased resulting in higher jacking loads [49].

Pipes and joints for installation by pipejacking and microtunnelling are designed for both gravity and pressure applications. The basic choice of pipe material is very much the same as it is for pipes for traditional trenching. Pipes are manufactured from:

(i) Concrete
- reinforced
- unreinforced
- composite concrete
- polymer concrete

(ii) Glass reinforced plastic (GRP)
(iii) Asbestos fibre and asbestos-free fibre cement
(iv) Vitrified clay
(v) Plastic
(vi) Ductile iron

8.1.2 Concrete pipes

Concrete pipes come in a number of forms but reinforced concrete is the material most commonly used for jacked installations for gravity sewers.

Reinforced concrete (RC). Pipes are normally factory-produced by the pipe-spinning method or vertically cast with high-frequency vibration. In some cases pipe casting yards have been set up at the site of the job. RC pipes come in sizes DN 250-5000 mm.

Joints for concrete pipes. These fall into two types: rebated, and collared or sleeved. Rebated (or scarfed) joints form a socket-and-spigot type of joint within the pipe wall. The sealing ring is set in the centre of the pipe wall. Rolling and confined sealing rings are available as shown in Figure 8.1.

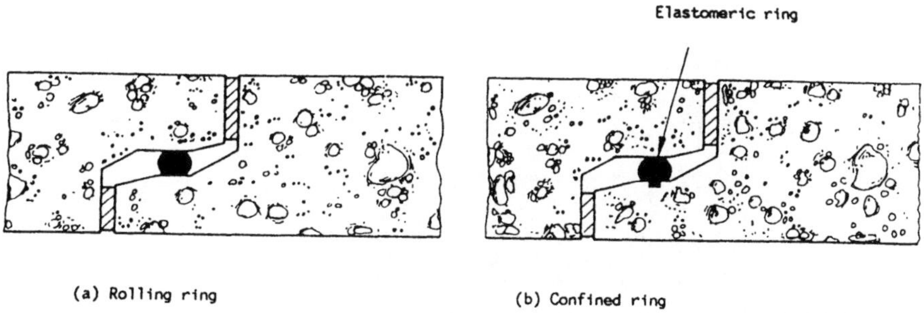

Figure 8.1 Rebated joints

Collared or sleeved joints (also termed butt joints) use the main body of the pipe as the spigot with a sleeve of another material. The sealing rings set in a recess on the exterior of the pipe. Joints can be of the fixed collar type with rolling or confined ring, or of the loose collar type with a confined ring, as shown in Figure 8.2

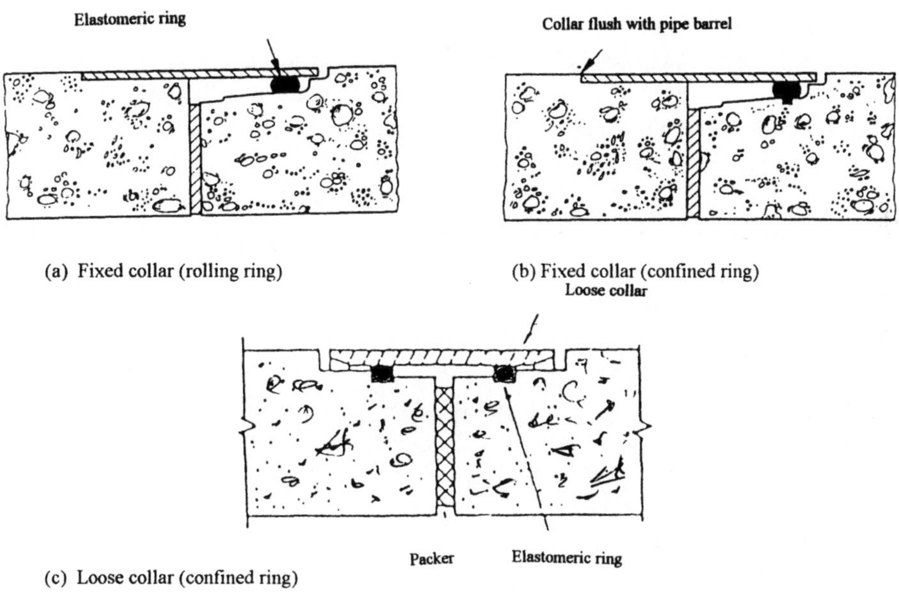

Figure 8.2 Collared joints (three types)

The rebated joint limits the axial load that can be applied to it because only approximately half the pipe wall thickness can truly be considered as load bearing. Failures caused by uneven loading can occur across the joint. Because of these limitations, the butt joint is now the most commonly used type of joint worldwide. In Europe, the loose collar finds some use in microtunnelling sizes, whereas the fixed collar type is predominantly used in sizes for pipejacking. A variation of the loose collar ring, shown in Figure 8.3, is used in Japan. This collar has a T cross section with the down part of the T acting as a thrust transfer ring.

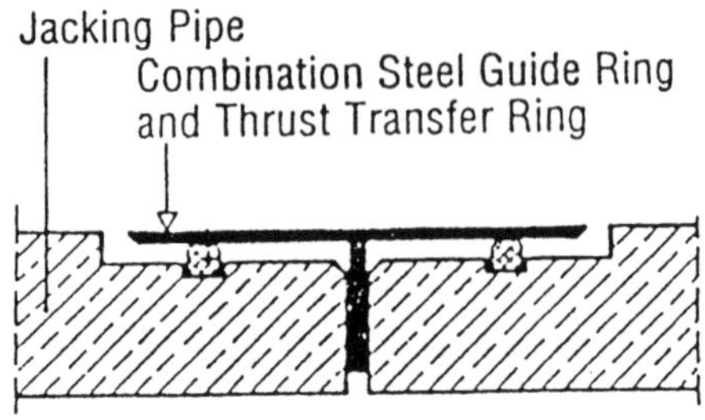

Figure 8.3 Loose ring joint (Japanese)

Sleeves need to be adequately protected against corrosion. Collars are made from a variety of materials including weldable steel, stainless steel and GRP, however not all materials are available throughout the size range. The thickness of steel sleeves varies with diameter, but there is some concern about the durability of steel buried in the soil. Consequently, various protective coatings are applied to the steel and some authorities call for a secondary sealing gasket. Where the corrosivity of the soil is high, a more corrosion-resistant collar is advisable.

It should be noted that Japanese manufacturers do not seem to be so concerned with this problem, as collars of mild steel with minimal protection are widely used. The PJ stainless steel joint shown in Figure 8.4 is an exception. The seal, fixed to the ribbed collar and made of a material which swells, covers the whole length of the joint.

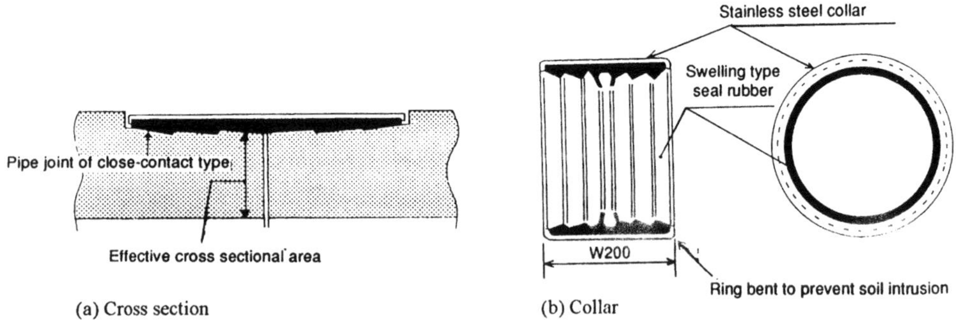

Figure 8.4 PJ stainless steel (drawing by Chichubu Concrete Industries Co. Ltd, Tokyo)

CS pipe from Japan has a modified joint and collar which is claimed to provide much greater flexibility and an ability to accept eccentric loading. See Figure 8.5.

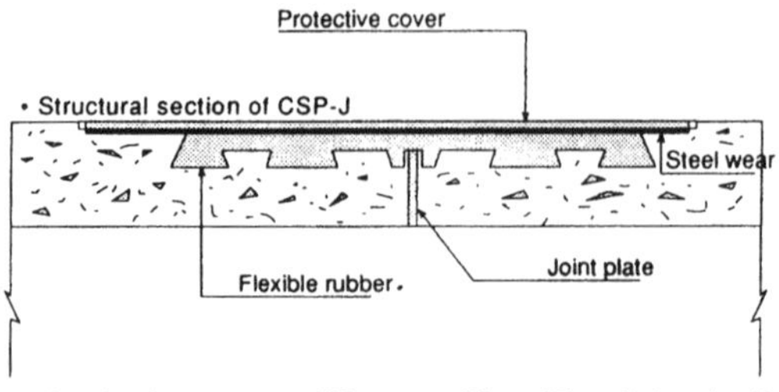

Figure 8.5 CS pipe joint (drawing by courtesy of Nagagawa Hume Pipes Industries Co. Ltd., Japan)

Table 8.1, which is based on Japanese practice, indicates the allowable jacking forces on a range of microtunnelling and pipejacking diameters. These jacking forces are somewhat higher than would be allowed using German guidelines. The allowable Japanese stress in concrete is 13 N/mm² whereas the German is 11 N/mm².

Table 8.1 Allowable jacking forces on concrete pipes (Japan)

Nominal ID (mm)	OD (mm)	Effective area (m²)	Allowable jacking force-tonnes
Microtunnelling diameters			
250	360	.0364	47.3
300	414	.0451	58.6
400	526	.0675	87.8
500	640	.0958	124.6
600	712	.1021	132.7
700	832	.1433	186.81
800	942	.1766	229.6
Pipejacking diameters			
900	1062	.2297	298.6
1000	1182	.2897	376.6
1100	1292	.3365	437.5
1200	1412	.4084	530.9
1350	1576	.4800	624.0
1500	1756	.6107	723.9

National standards. In Germany the relevant standard is DIN 4035, in UK BS 5911 and others exist in countries like Switzerland. The European Community (EC) and European Free Trade Association (EFTA) countries have agreed to produce unified mandatory standards. These are being prepared by the Comité Europêan de Normalisation (CEN); concrete pipes for jacking come under CEN Technical Committee 165, which has issued some draft standards and is preparing others.

In Japan the government body, The Japanese Sewage Works Agency, has issued a specification for jacking pipe, JSWAS-A2. However, in the USA, no ASTM standard for jacking pipe exists (as at 1993). The lack of a standard and variable pipe quality have inhibited the more extensive use of concrete jacking pipes.

Unreinforced concrete pipes. Unreinforced concrete pipes can be used for microtunnelling, mainly for DN 300 mm or less. One use has been for laterals in DN 100 and 150 mm. As these pipes are made about 20% thicker than reinforced versions, they are heavier to handle. Some problems have been noted with breakdown of the pipe structure under eccentric loading.

ARC Concrete Ltd and Yorkshire Water Plc have developed in the UK an interesting example of unreinforced concrete pipe. The pipes are manufactured by the centrifugal dry-spun process incorporating low frequency high amplitude vibration, which produces a very dense high-strength concrete to close tolerances. These are made in one size, DN 600 mm.

The joint for this pipe, shown in Figure 8.6, has some original features. The T-shaped steel collar is of the loose type with confined sealing rings. The down leg of the ring, which acts as a load transfer ring, is lozenge-shaped to concentrate the maximum stress onto the *middle third* of the section. The whole of the collar is encapsulated in a plastic coating.

Figure 8.6 Loose collar joint and pipe (drawing and photograph by courtesy of ARC Ltd.)

Concrete composite pipes. Superline pipe, from a process licensed from ARC, is one of the best-known composite pipes and one much used for jacking in Japan. Inner and outer faces are strengthened with a glass-reinforced cement (GRC) coating. The pipe construction and joint is shown in Figure 8.7. An allowable working stress of 22 N/mm^2 gives the same jacking load capability with a wall thinner than for RC pipe.

Another example of a composite pipe is manufactured by the spun cast process. This has a core of concrete with inner and outer layers of glass-fibre-reinforced polyester resin. Diameters available are DN 400 mm and larger.

Polymer concrete pipe. Its impermeability, high tensile and compressive strengths together with high resistance to corrosion make polymer concrete an attractive material. However, the high cost of these pipes, which can be several times that of an ordinary concrete version, is a deterrent to their greater use.

Polymer concrete linings are not strictly pipes, but can be placed in situ to form a *pipe* direct in the ground. A microtunneller or TBM bores a tunnel and at the same time pulls behind it a sliding shutter. Polymer concrete from a mixing plant is pumped into the shutter to form a continuous lining with no joints. It is possible to control the setting time of the polymer to provide early strength and give a jacking reaction. Most of this work has been done in large diameters by German companies. For this type of lining, a remote-controlled system for small diameters, called the M2 method, has been used by NTT on a trial section in Japan [50].

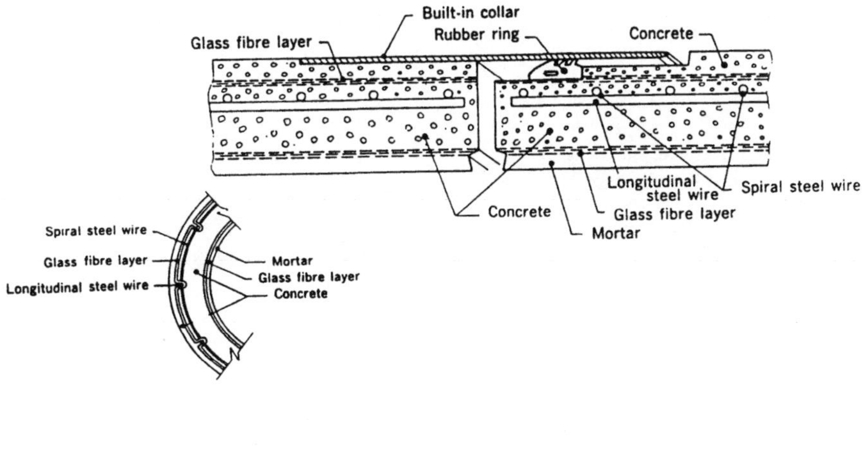

Figure 8.7 Superline joint (drawing by courtesy of Japan Superline Pipe Association)

Concrete pipes for pressure mains. The concrete pipes and joints discussed so far are designed for gravity sewers. Joints have to be watertight but need to have only a limited capability of withstanding internal pressure. Typical specifications for sewer pipe will require a pipe to be tested to a 14 m head. However, with careful detailing it is possible to obtain higher test capability. For a pipejacking installation in London, pipes were tested to 30 m head.

Prestressed cylinder pipes, used for water transmission mains have been adapted for use with microtunnelling and jacking. A number of variations on joint configuration have been developed, of which three examples from the French company, Bonna [51] are shown in Figure 8.8. The minimum diameter for this type of pipe has to date been DN 800 mm.

8.1.3 Glass-reinforced plastic pipes (GRP)

GRP pipes are manufactured by one of two processes: by centrifugal casting and by filament winding. Whereas the filament-wound variety has been little used in jacked applications, centrifugal cast fibre-glass-reinforced plastic matrix pipes (FRPM) have been extensively used since the early 1980s in Europe, Japan and North America on sewer installations by pipejacking and microtunnelling. The lack of suitable concrete pipe has made FRPM probably the most common pipe for microtunnelling in the USA.

The best-known manufacturer, Hobas, produces pipes in diameters from DN 200 mm to DN 2400 mm. Wall thicknesses can be from 30 to 60 mm. Table 8.2 shows the dimensions and safe jacking loads on Hobas pipes. Jointing is made by a GRP collar glued onto the recessed trailing end. A restrained sealing ring is incorporated to ensure watertightness. The recessed spigot end, which is pushed into this collar, also incorporates a restrained sealing ring. Figure 8.9 gives an example of this form of joint, while Figure 8.10 shows a pipe being installed by microtunnelling.

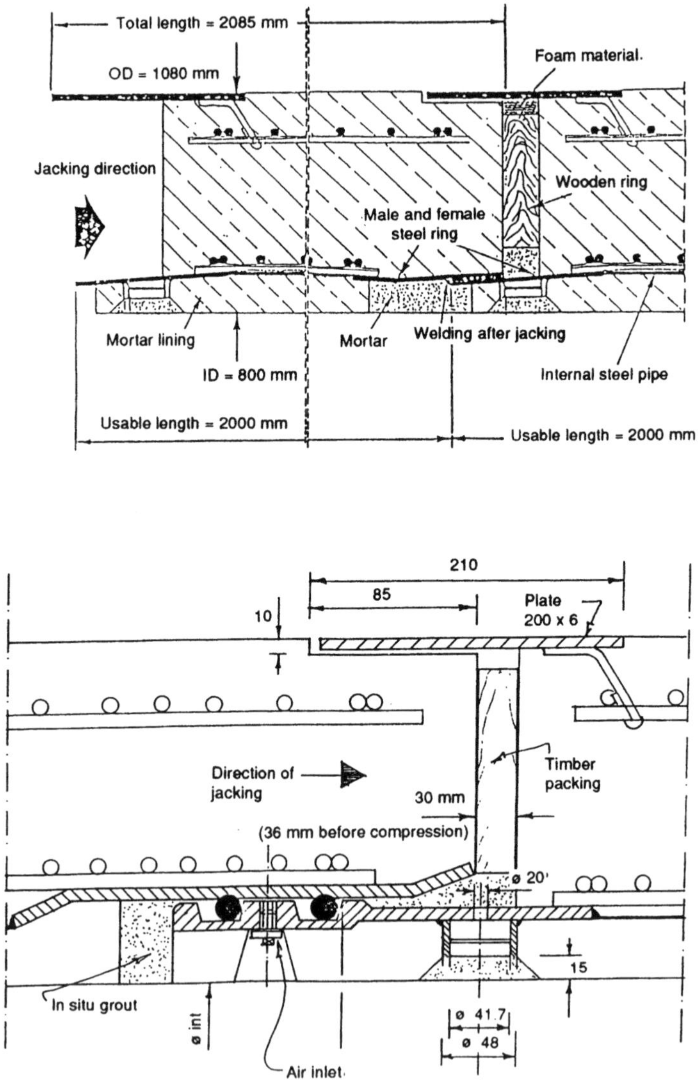

Figure 8.8 Bonna pressure pipe joints

Table 8.2 Dimensions and safe jacking loads on Hobas pipe (source, Hobas Durotec GmbH)

DN	Wall thickness (mm)	Safe load (ton)	Wall thickness (mm)	Safe load (ton)	Wall thickness (mm)	Safe load (ton)
272	40	34				
324	40	42				
401	40	54	50	66		
501	40	69	50	85		
616	40	86	50	107	60	128
718	40	101	50	127	60	152
752	40	106	50	133	60	160
820	40	117	50	146	60	176
924	40	132	50	166	60	200
1026	40	148	50	185	60	223
1229	40	178	50	224	60	271
1439			50	264	60	319
1638			50	302	60	365
1942					60	413
2047					60	460

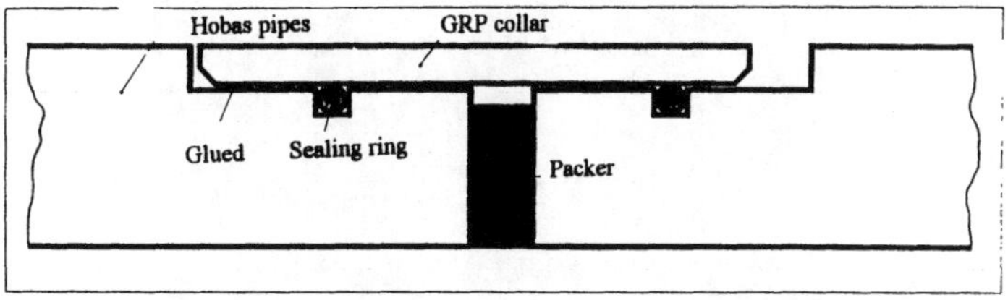

Figure 8.9 Hobas joint

Ground conditions should be considered before selecting FRPM pipes as they can be susceptible to damage by boulders or cobbles in the bedding. For any GRP pipe, it is also important to ensure that the ground is not contaminated by solvents which could attack the material. After they had been jacked at a job in Staten Island, USA, DN 1500 mm FRPM pipes were found to have the collar sleeves coming away from the joints and to have been stacking up on the outside of the pipe. The joints, of the push-together elastomeric gasket-sealed bell-and-spigot type, have normally behaved well. The possibility is that the joint design did not stand up to working at greater depths (27 m) with a high water table without allowing soil to be forced into the joint. Subsequently the joint was modified to eliminate any gap [52].

FRPM pipes in sizes from DN 500 to 800 mm with wall thicknesses of 18 to 20 mm have been installed in Japan by the LLB Gripper method [53], described in Chapter 5.

FRPM Pipes for pressure mains. FRPM pipes have been installed by traditional trenching work for pressure applications such as pumping mains. In theory it should be possible to use such pipes for operating pressures up to, say, 10 or 15 bar. However no authoritative design guidance or experience is available for their use in pressure applications.

Figure 8.10 Hobas pipe (photograph)

8.1.4 Asbestos cement pipes

Asbestos cement pipe (covered by ISO standards 160, 881 and 2785) is likely to be phased out for general use because of the health hazards associated with asbestos. Having invested some $100 million in researching an alternative fibre, Eternit, the leading manufacturer, is now producing asbestos-free fibre cement pipes at its companies in Germany, Belgium and Switzerland.

For trenchless applications, the onetime asbestos pipe, now to be asbestos-free fibre cement, is used in Europe for sewer installation by microtunnelling and to a small extent by pipejacking. It would appear that a substantial percentage of the small microtunnelling (DN 250 - 400 mm) jobs in Germany employ fibre cement pipe.

8.1.5 Clay pipes

Clay pipes have become well-established for trenchless installations in most European countries and to a lesser extent in Japan. They have also been introduced in the USA. They have some inherent advantages, most importantly:

- High chemical resistance with no need for secondary lining.
- High crushing and beam strengths and can take high jacking loads.

Improved manufacturing techniques, including machined pipe ends, have overcome traditional problems of dimensional tolerance and lack of straightness. However, clay pipe needs careful handling as it is brittle and therefore susceptible to fracture.

Pipe sections can be obtained in 750 mm, 1000 mm and 1250 mm lengths for smaller diameters, and 1.5 m and 2 m lengths for larger diameters. Table 8.3 sets out the dimensions and safe jacking loads for the Steinzeug (Germany) range of clay pipes.

Table 8.3 Dimensions and jacking loads for clay pipes (after Steinzeug GmbH)

DN (mm)	O.D. (mm)	Wall thickness (mm)	Jacking load (kN)
TYPE 1			
150	207	28	131
TYPE 2			
200	276	38	212
250	355	52	529
300	406	52	624
400	556	78	1315
500	658	78	1571
TYPE 3			
600	760	80	1609
700	862	80	1675
800	970	85	1982
1000	1178	88	2070

Collar arrangements vary, but Figure 8.11 shows three joint configurations used in Germany:

- DN 150 mm, a polypropylene body with restrained elastomer seal.
- DN 200 to 400 mm, a steel profiled ring with restrained elastomer seal.
- DN 500 to 1000 mm, a stainless steel ring with restrained elastomer seal.

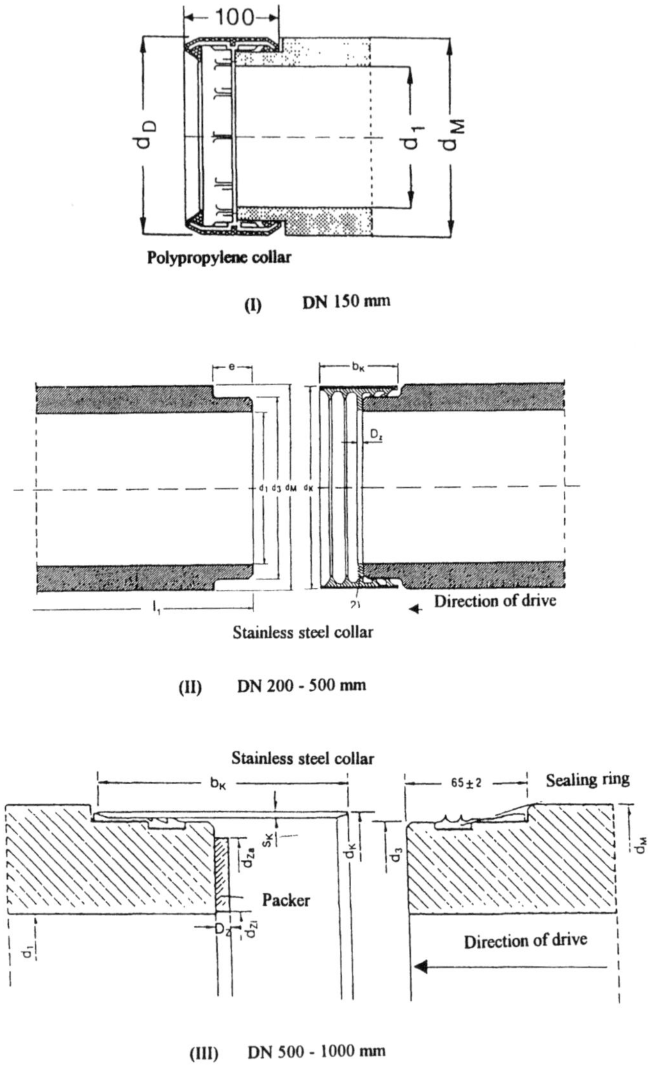

Figure 8.11 Clay pipe joints (drawing by courtesy of Steinzeug GmbH)

Although available in diameters up to 1000 mm, the smaller sizes are the more commonly used. In Germany, clay pipes are popular for microtunnelling small sewers and house connections in DN 350 mm, whereas in the UK and Japan preferred sizes are from DN 150 mm to 600 mm.

Present standards for vitrified clay pipe, German DIN 1230 and British Standard 65 and those for other European countries will in time be superseded by European Standard EN295. The US standard for vitrified clay pipe is ASTM C 1208-91.

8.1.6 Plastic pipes

Japan has taken the lead in using PVC for direct jacked installation. With a microtunnelling technique, the Japanese Envi-Liner [54] directly installs PVC pipes in DN 200 mm to 300 mm. Figure 8.12 shows some PVC pipes for microtunnelling.

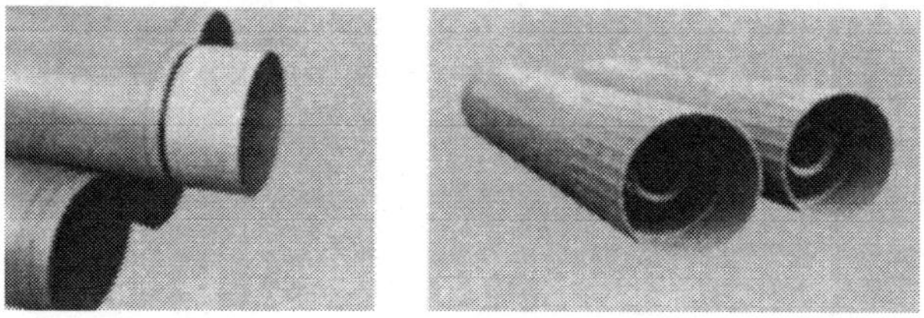

Figure 8.12 Plastic pipe (Japanese)

Two questions about PVC pipe present themselves. One is whether the side support needed for PVC (as for all plastic pipe) can be satisfactorily achieved in a trenchless situation. This problem has been the subject of some recent testing in Japan [55]. The other is whether a satisfactory flush joint can be achieved with thin-walled PVC pipe.

An interesting new PVC pipe joint has been developed by Carlon, USA, for the Vylon pipe. Vylon pipe uses a profiled wall which provides an "I-beam" section. This was used in *gripper* installation trials conducted by the Trenchless Excavation Centre (TEC) [28]. The joint for a microtunnelled installation is made within the outer *flanges* of the wall and is shown in Figure 8.13. A fibreglass insert ring with a multi-fin gasket provides the seal. Pipes of this type can be made in larger diameters, such as the DN 600 mm used in the trials.

8.1.7 Ductile iron - pressure applications

The greatest application for ductile iron (DI) has been for the pipejacking and microtunnelling installation of pressure pipe in Japan. Singapore has also installed DI pipe for sewage pumping mains and for industrial water lines. In Jakarta, Indonesia, DI mains have been installed by both pipejacking and microtunnelling.

DI pipes are available in a range of diameters from DN 300 mm to DN 2600 mm in standard sections 4 m long. Pipes are available in four classes. Figure 8.14 illustrates the joint configuration for Kubota pipes, which is formed by a low-profile socket and spigot with mechanical stud-bolt and nut connecting an external flange. In order to achieve a uniform flush external surface, a layer of reinforced concrete is added. It is also standard practice to provide DI pipe with a cement mortar lining.

The allowable jacking load in tonnes is estimated for Class I pipes to be 75% of the diameter in millimetres. A DN 1200 mm pipe will take 900 tonnes, which is substantially greater than for other forms of jacking pipe. Special pipes are also available to permit the use of IJSs for longer drives. In the UK, DI jacking pipe is available in a limited range of diameters. Its greater use has been discouraged by its standard length of 4 m and its high cost.

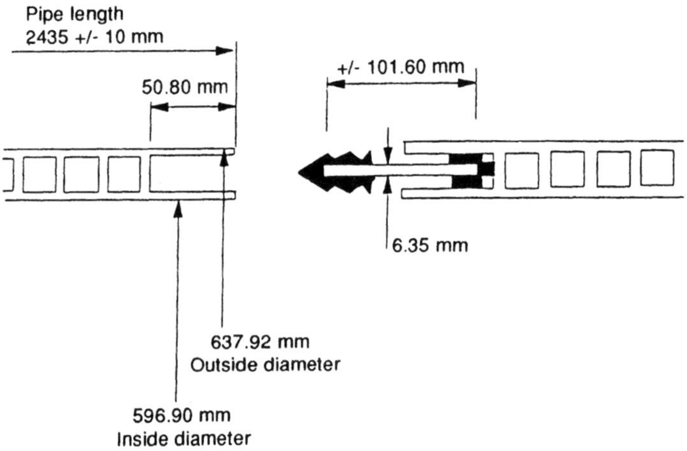

Figure 8.13 PVC pipe wall section - Vylon pipe (drawing by courtesy of Carlon, USA)

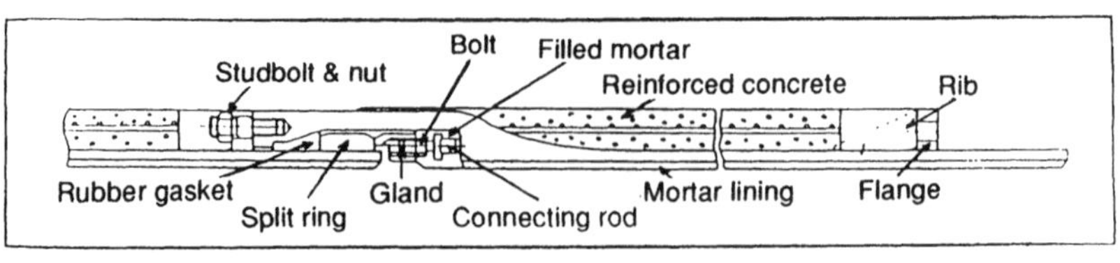

Figure 8.14 Kubota ductile iron pipe (drawing by courtesy of Kubota)

8.1.8 Steel pipes

Steel pipe with site-welded joints is generally used as a jacked casing, particularly at crossings. Although some casing has been installed by microtunnelling, it is more usual for steel pipe to be used in man-entry diameters. As welding joints at site is time-consuming, pipe sections in lengths up to 6 m or more are preferred, but for these much longer drive shafts must be provided.

If steel pipe is to be installed as a permanent service pipe, the exterior must be protected in a way which prevents it being damaged as it is jacked through the ground. This can be done, but the technical and cost issues discourage its general use.

8.2 Structural design principles and methods

8.2.1 Principles of pipe analysis

The structural design of pipes must be carried out for both the installation and the service conditions, the loading conditions for which differ considerably. The loads on pipes during installation arise primarily from the jacking forces applied by the rams housed in the jacking pit and IJSs, and the steering jacks located in or near the shield. These loads are related to the face resistance and line friction loads. These forces result in complex patterns of primarily axial stress within the pipe wall, the pattern being further complicated, and dominated, by relative misalignment of the pipes in the line. This subject is discussed in detail in Section 10.1.

The stresses perpendicular to the pipe axis during installation are generally localised, in that they act on one side of the pipe only and reflect the local relative pipe alignment conditions. The exceptions to this case include pipes being installed in:

(i) Dense granular materials with minimal or no overcut and no facility for stress relief (such as full earth pressure balance) at the face.
(ii) Heavily overconsolidated clays that swell due to stress relief, remove the effects of overcut and *squeeze* the pipe.
(iii) Granular materials (especially below the water table) or weak alluvial soils, which collapse immediately and completely onto the pipes.
(iv) Rock where dilatant rock flour fills the void created by the overcut and similarly causes the pipe to be *squeezed* by the ground.

Pipes being *squeezed* by the ground (cases ii and iv above) should be analysed carefully; engineering judgement and specialist geotechnical advice will be necessary to determine the likely pipe loads. In other cases the loading condition is unlikely to be more severe than that experienced in service and the analyses described below will be applicable.

The most significant loads on buried pipes in service are those acting perpendicularly to the pipe axis, and derive primarily from the soil overburden, surface dead and live loads, and internal and external fluid pressures. These loads are discussed in detail in subsection *8.2.2*. The loading experienced by the pipes is greatly affected by the stiffness of the pipe, or more particularly the relative stiffnesses of the pipe and the soil, and can vary significantly with time as ground conditions change after pipe installation.

Pipes have traditionally been categorised as either rigid or flexible, and designs have been carried out accordingly. Rigid pipes resist applied loads by bending action in the pipe walls, the loads being subsequently transferred to the underlying soil through a good quality bedding material (Figure 8.15 a). For rigid pipe design, therefore, the support condition below the pipe, and in particular the width of the support, is of primary importance. Overstressing of the pipe will result in fracture of the pipe wall at relatively small pipe wall deflections, and consequently little deformation of the pipe occurs under load.

By contrast a flexible pipe is designed to deform considerably under load without fracture and resists applied load by generating passive soil resistance on the side walls of the pipe. In this case, therefore, the structural unit consists of both the pipe and the surrounding soil, and the stiffness of the surrounding soil becomes of prime importance to the pipe stability. The structural action is more complex than this, however, since the process of pipe deformation generates arching in the soil overlying the pipe and thus load shedding to the masses of soil to either side of the pipe, the *sidefill*. The combined action is shown in Figure 8.15 b, which emphasises the double effect that the stiffness of the sidefill has on the resistance to load, and thus on the magnitude of pipe deformation.

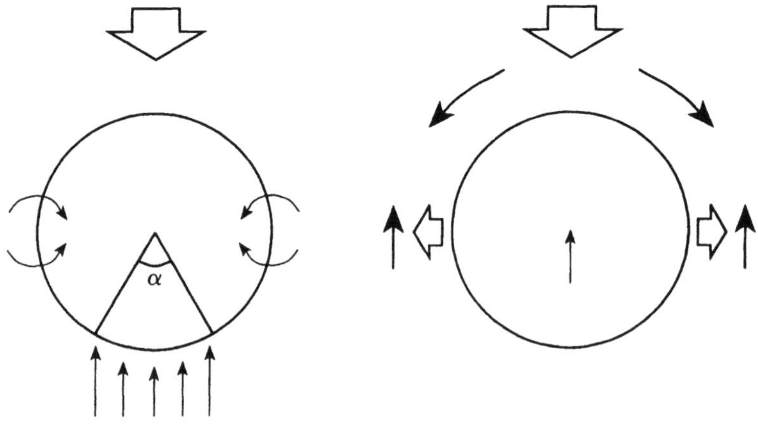

(a) Rigid pipe (b) Flexible pipe

Figure 8.15 Structural action of rigid and flexible pipes

However, these categories of rigid and flexible pipe should be considered as relative only as they have been blurred by the advent of different forms of pipe. For example, the structural capacity of a concrete sewer pipe can be significantly increased when installed in a thoroughly compacted, good quality sidefill.

In terms of pipejacking and microtunnelling, and with the caveat mentioned above, pipes manufactured from concrete, clay, steel-concrete composites and ductile iron can be considered in most cases to be rigid, whereas steel pipes (which typically have a relatively thin wall) can be considered to act as a flexible pipe. A complete analysis of soil loads on buried pipes was first reported by Marston [56] in which the concept of relative rigidity was effectively, though not explicitly, introduced. Spangler [57] further developed these ideas with specific application to flexible pipes. The concept states that an inclusion (pipe) that is stiffer, by virtue of its own stiffness or its restraint, than the surrounding medium (soil) that is free to undergo movement (settle) will attract a greater proportion of vertical load than if there were no difference in stiffness between the two. The opposite (i.e. a reduction in load) would occur if the inclusion was less stiff than the surrounding medium. Thus a rigid pipe constructed on a rigid bedding in an embankment that is settling with time will experience a load greater than the free-field vertical stress (a case termed the positive projection condition), whereas a relatively flexible culvert under the same condition will experience a load less than the free-field vertical stress.

The method by which the pipes are installed will also influence the loads that will act on them. Since trenching is the traditional method of pipe installation, most work on the subject compares the loads acting on pipes laid by trenchless methods with those laid by open cut. O'Rourke and others [58] describe the soil-load transfer mechanisms in Figure 8.16. For a pipe buried in a trench, the backfill will settle relative to the undisturbed material that constitutes the trench wall. This will result in shear forces acting to resist the relative movement between the backfill and the trench wall, thus reducing the load that will be transmitted to the pipe. In effect this relative settlement is inducing arching and load shedding. The load acting on the pipe will therefore be less than the weight of the rectangular soil prism directly overlying the pipe.

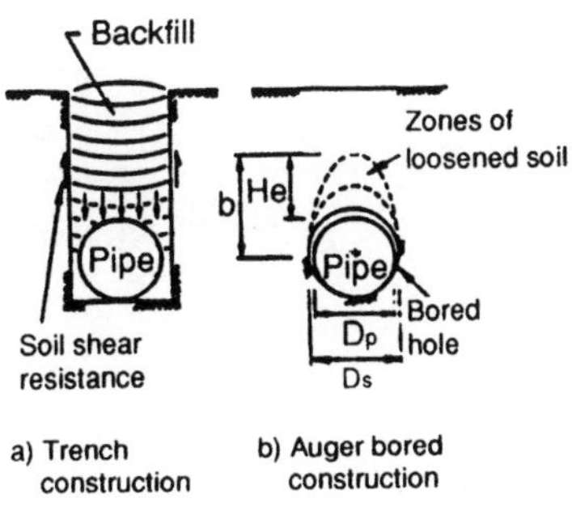

Figure 8.16 Soil load transfer mechanisms

For jacked pipes following a shield of larger diameter, the overcut will cause zones of loosened soil to be generated once the soil collapses onto the pipe at some time after excavation (a topic that is further discussed in Sections 9.2 and 10.1). Unless the soil is a very loose sand or a very weak alluvial material, the zones of loosened soil will result in a height of soil (H_e) that is much less than the depth of cover and thus a soil load that is much less than the prism load.

It can be shown that, assuming zero volume loss at the face, the volume of soil loading the pipe (V_L) is

- eqn 8.1

$$V_L = \frac{\pi D_s b}{4} + \frac{\pi D_s^2}{8} - \frac{\pi D_p^2}{4}$$

where V_L = Volume of soil contained within zone of loosened soil (m³)
D_s = Diameter of shield (m)
D_p = Diameter of pipe (m), and
b = Distance from pipe axis to top of zone of loosened soil (m). See Figure 8.16.

However since the overcut is usually relatively small in comparison with the pipe diameter, a sufficiently accurate estimate of pipe load can be gained from

- eqn 8.2

$$W_s = \gamma_b \left(\frac{\pi D_s b}{4} - \frac{\pi D_s^2}{8} \right) = \frac{\pi \gamma_b D_s}{8} (2b - D_s)$$

where W_s = Weight of soil acting on pipe (kN), and
γ_b = Bulk unit weight of soil (kN/m³)

The difficulty in the case of this equation is in the choice of the parameter b. Research [59] has indicated that the loosened zone extends vertically to a maximum of 2-3 pipe diameters in dense sand and that the zone is likely to extend to the soil surface in loose sand. Thus in a dense sand the maximum load on the pipe is likely to be

- eqn 8.3

$$W_s = 0.75 \pi \gamma_b D_s^2$$

whereas in loose sand the full soil prism load is likely to act, in which

- eqn 8.4

$$W_s = \gamma_b H D_p$$

where γ_b = Bulk unit weight of soil (kN/m³)
H = Cover depth (m), and
D_p = Diameter of pipe (m)

Pipe loads in heavily overconsolidated clay soils might be expected to be similar to those in dense sand, and pipe loads in lightly overconsolidated or normally consolidated clay soils similar to those in loose sand, although no direct experimental data are currently available to substantiate this observation. O'Rourke and his colleagues [58] produce a graph of the normalised height of loosened soil against bulking factor for different percentages of volume loss, although the ranges quoted for particular soil types take no account of soil density (sands) or overconsolidation ratio (clays) and, indeed, assume that bulking will occur. In very loose (e.g. aeolian) sands and soft, weak, young alluvial soils contraction rather than dilation (bulking) would be expected. However, this work relates to auger boring and relatively large overcut ratios (typically 1.1) are quoted.

Using a combination of Marston's [56] theory and equations for the volume of soil in the loosened zone, they produced a graph indicating the ratio of elliptical collapse (C_{de}) to ditch conduit (C_{dM}) load coefficients for different cover depths and bulking factors (Figure 8.17). This graph shows that close to the surface the coefficients are similar, and will be close to the soil prism load, since H_e will equal the depth of cover (H) and there is little opportunity for trench wall friction to relieve the backfill load on the pipe. At greater depths both factors will indicate relief from the prism load, but the elliptical collapse coefficient (C_{de}) reduces faster than the ditch conduit coefficient (C_{dM}). Thus at a cover to pipe diameter ratio of 10, the loads based on the elliptical collapse mechanism are approximately 20% to 55% of those predicted by the ditch conduit method, which are in turn considerably smaller than the soil prism load.

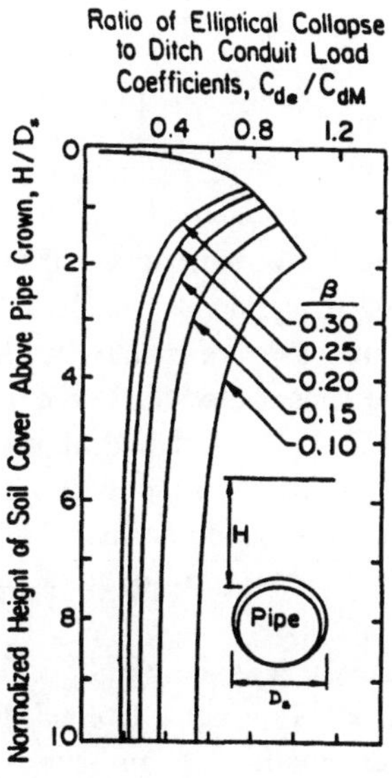

Figure 8.17 Ratio of soil load coefficients for different depths of cover and bulking factors (after O'Rourke et al)

Certain construction factors can also influence load transfer onto pipes. Firstly the use of minimal or no overcut to limit settlement, particularly in conjunction with earth pressure balance machines, could result in little or no soil loosening and full load transfer from the soil onto the pipe. The magnitude of load applied to the pipe in this case would then depend upon the relative rigidity of the pipe to the soil. A rigid pipe would attract a load equal to or greater than the prism load depending on subsequent surrounding soil settlement. Where earth pressure balance is not applied, stress relief at the face will probably result in some loosening above the pipe and a somewhat less onerous loading condition.

The same load transfer mechanism is valid for cases where overcut is used and the annular gap is immediately filled with a lubricant under a pressure (measured at the point of injection) equal to the weight of the soil trying to collapse onto the pipe, and that is both relatively incompressible and cannot subsequently disperse into soil.

Conversely the use of a large overcut might result in the breakdown of arching mechanisms and total collapse of soil above the pipeline, although this has not been found in laboratory work [60]. Overpressurisation of slurry during drives in clay could also result in longer-term consolidation of clay soils as high pore water pressures during the drive dissipate with time, and therefore increase soil loading on pipes.

Finally in drives in granular soils below the water table, and in particular in cases where water flow towards the pipeline can occur, the collapse of soil onto the pipe will be immediate and more importantly the arching mechanisms, and thus the elliptical collapse theory, could break down.

These few examples show how important it is to consider both the construction operations and the geotechnical conditions before adopting the results of theoretical solutions to the calculation of loads on pipes.

8.2.2 Loads on buried pipes in service

When designing pipes for pipejacking, a number of loading conditions and construction parameters must be examined. These are exhaustively listed in the German standard (ATV 1990), the most comprehensive worldwide standard [61]. The loads acting on jacking pipes perpendicularly to the pipe axis are summarised in Table 8.4 [21]. This is a useful checklist, although it should be remembered that these loads will vary with time and the worst combination of maximum likely loads should be used for pipe design. A good example of this variation concerns soil loads on pipes jacked into an oversized void in clay. Initially the clay will be self-supporting and no load will be applied to the pipe. As the soil equilibrates the load will increase up to some maximum level well after construction has been completed.

Certain loads are easy to determine, such as the weight of the pipe and the fluid that it is intended to carry when the pipe is full. Excess internal pressures both positive and negative, for example water or gas distribution pipe pressures, are also well defined from experience of operating conditions. The effect of internal pressures will be reduced if there is an external hydrostatic pressure due to the pipe being below the ground water table (or increased if combined with negative internal pressures) and can be calculated to provide the worst case all-round loading. For this, it is usually adequate to calculate the pressures at the depth of the pipe axis and consider them to be uniformly distributed around the pipe circumference. However, for certain cases (e.g. large diameter pipes) the difference in pressure between the top and bottom of the pipe should be taken into consideration.

Table 8.4 Loads acting perpendicularly to the pipe axis, after ATV (Stein)

Load	During Construction	In Service
Pipe dead weight	√	√
Vertical soil load	√	√
Lateral soil pressure	√	√
Traffic loads	√	√
Uniformly distributed surcharge	√	√
Strip load	√	√
Point load	√	√
Internal fluid (pipe full) load		√
Excess internal pressure		√
Internal negative pressure		√
External hydrostatic pressure	√	√
Supporting (e.g. lubricating fluid	√	
pressure)	√	
Compressed air	√	√
Uplift (or buoyancy) pressures		

Where the pipe is being designed for vertical load-carrying capacity, the lowest likely water table position should be used to calculate the hydrostatic stress. However, for buckling calculations the highest likely water table should be used. The temporary construction loads due to compressed air and supporting pressures, such as lubricating fluid pressure, can usually be predetermined. The values can then be quoted to the contractor as safe working limits. Uplift, or buoyancy, pressures act to relieve the pipe of load and are usually ignored in design. In certain cases accurate estimations of the soil load on buried pipes are possible using the equations propounded.

The ATV method. It has been stated earlier that the first approximation to the soil load concerns the soil prism effect (equation 8.4), which should be modified if the pipe is below the water table to

$$W_s = (\gamma_b . H - \gamma_w h_w) D_p \qquad \text{- eqn 8.5}$$

where W_s = Weight of soil acting on pipe (kN)

γ_b = Bulk unit weight of soil (kN/m³)

H = Depth of cover (m)

γ_w = Unit weight of water = 9.81 kN/m³

h_w = Height of water table above pipe axis (m), and

D_p = Diameter of pipe (m)

For this calculation the lowest height of water table should be used, since the water pressure will reduce the soil load. The ATV (1990) calculation method uses a form of ditch conduit model (termed the *silo-theory*) based on the observations of Terzaghi during his famous trap-door experiments. The basis of the silo theory, seen in Figure 8.18, considers a prism of soil of height H and width b_s that settles relative to the soil at the side of the prism. This results in a relieving effect due to frictional effects on the vertical sides of the prism, as in the ditch conduit model. For this relieving stress to occur in full, the vertical movement must be greater than a minimum value. The overcut required to ensure that such movements occur could be unacceptably large, since, by definition, there would necessarily be associated surface settlement.

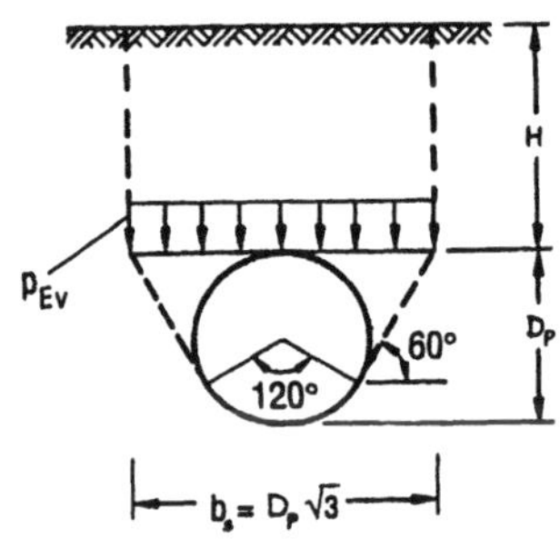

Figure 8.18 Basis of the Silo Theory Model to calculate soil load acting on buried pipes

Several researchers have used different widths of soil prism to determine the soil loading on pipes. The widths range from the pipe diameter (see the ditch conduit model shown in Figure 8.16) to that shown in Figure 8.18. Whichever width is used, the relief in stress on the prism (σ_r) is calculated using

$$\sigma_r = p_v K \tan \delta \qquad \text{- eqn 8.6}$$

where p_v = Free-field vertical stress (kN/m^2)
K = Coefficient of lateral earth pressure
δ = Coefficient of friction between soil prism and side walls of surrounding soil (degrees)

The value of δ recommended by Terzaghi is the angle of internal friction of the soil (Φ'), whereas in the German standard a more conservative value of $\Phi'/2$ is used. The value of K recommended by ATV is 0.5, which is low for dense sands and heavily overconsolidated clays but is once again generally conservative.

The ATV (1990) method thus determines a uniform surcharge acting on the pipe using

$$P_{Ev} = \kappa \gamma_b H \qquad \text{- eqn 8.7}$$

$$\text{where } \kappa = \frac{1 - e^{-2K\tan(\Phi'/2)\,(H/b_s)}}{2K\tan(\Phi'/2)\,.\,(H/b_s)} \qquad \text{- eqn 8.8}$$

where P_{Ev} = Equivalent vertical stress acting at the level of the pipe crown (kN/m²), and

κ = Stress reduction factor due to silo effect

A chart for determining the reduction factor is reproduced in Figure 8.19. This chart shows that the reduction in load is greater with both depth and increasing angle of internal friction of the soil, although the loads would still appear to be significantly greater than those advocated by O'Rourke [59]. It should be note here also that when below the water table, no reduction should be applied (ie $\kappa = 1.0$) and the fail soil prism load should be used for design. Once the vertical stress has been determined, the equivalent horizontal earth pressure acting at the level of the horizontal axis of the pipe (P_{Eh}) can be calculated using

$$P_{Eh} = (P_{Ev} + 0.5\, D_p\, \gamma_b)\, K \qquad \text{- eqn 8.9}$$

In using this equation, values of K ranging between 0.3 and 0.5 are recommended, depending on the soil type and whether the annulus is injected with lubricant or grout.

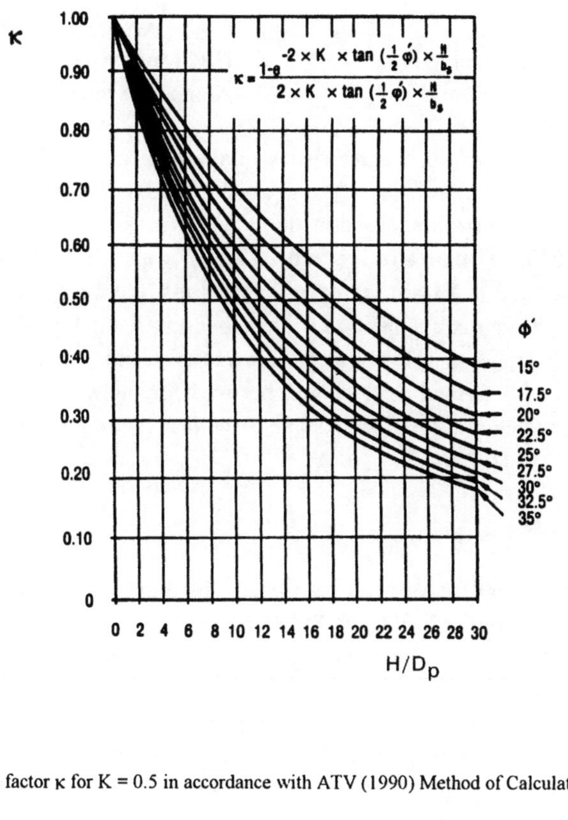

Figure 8.19 Reduction factor κ for K = 0.5 in accordance with ATV (1990) Method of Calculating Soil Loads (after Stein et al.).

Where relatively flexible pipes are used, the horizontal stress will increase above this value and this should be allowed for in the calculation of pipe stresses. Uniform surface surcharge loads should be added to the vertical soil loads calculated above. The load on a pipe due to strip loads and point loads can be calculated using traditional geotechnical principles, or by simple assumptions of attenuation of stress with depth, to give an equivalent additional pressure at the pipe crown.

Live loads. The loads due to surface traffic are complex to calculate with precision, although several established techniques are available. The loading for design is generally split into two components: the directly calculable load using static theory, and an impact factor, which is used to take account of the dynamic nature of the loading. In the UK an authoritative document by Young and O'Reilly [159] explains how to calculate the loads due to road traffic and railway carriages. Different combinations of loading are considered and in the case of railways, single and multi-track loadings are allowed for. The German approach, given in ATV (1990), covers aircraft loading in addition to traffic and railway loading. More recent research in the USA reported by Ingraffea et al [62] and Stewart and O'Rourke [63] on loads beneath highways and railroad crossings recommends calculation techniques which follow the classical American approach of determining the applied surface stress and then multiplying it by several factors to account for such parameters as:

(i) Impact factor
(ii) Highway/railway stiffness
(iii) Highway/railway geometry
(iv) Highway pavement type
(v) Vehicle axle configuration
(vi) Number of railway tracks

These parameters are linked to characteristics of the pipe and its installation (e.g. pipe size, pipe stiffness and depth of burial) and can be determined by reference to design charts. the simplest approach adopted in the USA is to assume that the axle loads are uniformly distributed over an area of 2.4 m by 6.1 m. Recent US research has shown that this approximate method gives an appropriate stress distribution for railway loadings at typical pipeline construction depths. The pipe loads beneath highway crossings were found to be approximately one half of that for railway crossings. An important finding was that impact factors of 1.65 and 1.60 should be used for highway and railway loadings respectively. These values agreed with some American standard guidelines, but were significantly greater than others.

8.2.3 Design of pipes

Once the various pipe loads have been calculated, the most severe combinations of load that could occur at any point during the life of the pipe must also be calculated, both in terms of vertical, out-of-balance loads for pipe structural capacity and maximum all-round loads for buckling capacity.

The design of pipes is beyond the scope of this book and in most cases is not required of the design engineer, since pipes with an appropriate structural capacity would be chosen once the loadings had been determined. The one exception to this rule concerns the case of flexible pipes, for which appropriate soil stiffness parameters require selection in design. Much experience has been gained with selection of soil stiffness (E′) parameters for buried pipe design, but the choice of E′ values for pipejacking and microtunnelling pipes is far more difficult due to the effects of local loosening of soil by overcutting. The best reference for guidance on this matter is O'Rourke et al [59], who have taken field measurements of deformation and have back-calculated E′ for these cases.

The equation for flexible pipe design has been developed by Spangler [57] and modified by Watkins and Spangler [64], which states that the horizontal deflection of a flexible pipe (Δx) is given by

$$\Delta x = \frac{D_L\, k_b\, W}{\dfrac{E_p I}{r_p^3} + 0.061\, E'} \qquad \text{- eqn 8.10}$$

where
D_L = Deflection lag factor
k_b = Bedding factor
W = Total load on pipe per metre run (kN/m)
E_pI = Bending stiffness of pipe ring (kNm)
r_p = Radius of pipe (m)

By assuming uniform support over a 90° arc of the base of the pipe and taking data that correspond to the end of primary pipe compression (i.e. D_L = 1.0), the predicted against the measured pipe deflections assuming E′ values of 3.5 and 7.0 MN/m² respectively were found to be those shown in Figure 8.20. These data indicate an E′ value of 3.5 MN/m² serves as an approximate lower bound to all measurements and can therefore be used in design with some degree of certainty. In particular cases E′ values can be adjusted with appropriate engineering judgement, although high values should not be used unless soil disturbance can be kept to a minimum during construction. Design equations for buckling and pipe wall stresses can then be used, and guidance sought from pipe manufacturers on their recommended design methods.

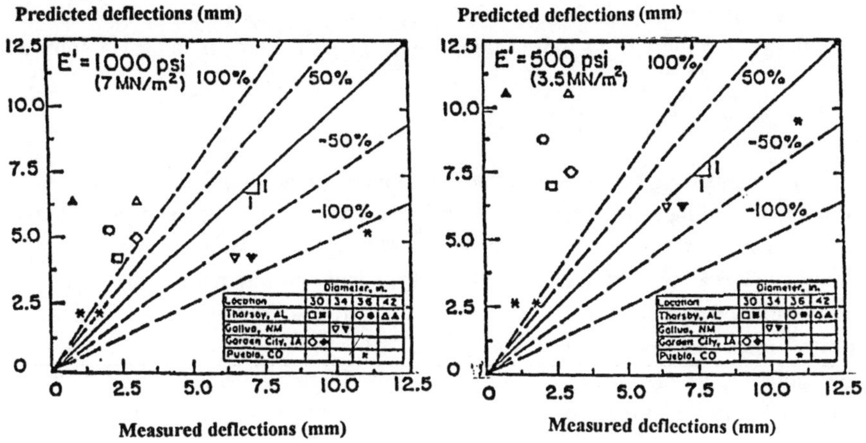

Figure 8.20 Predicted vs. measured pipe deflections for steel pipe installed by auger boring (after O'Rourke et al.).

Safety factors. Safety factors in the design should allow for uncertainties about:

(i) The design equations or assumptions.
(ii) Loading conditions.
(iii) Material properties.
(iv) An overall safety factor.

The best approach would be to take the worst likely conditions that could possibly occur and to factor the design up by a small amount (say by a factor of 1.1). This is far more certain than adopting average conditions and applying substantial safety factors to each element.

8.3 Protective linings and coatings

Some pipe materials, particularly concrete have poor resistance to chemical and biological attack in and around a sewer. Engineers, therefore, increasingly specify some form of protective liner for them.

8.3.1 Plastic sheet linings

The most common approach is to provide an internal lining of PVC or other plastic sheeting. This can be applied either during or after pipe manufacture.

Some Japanese manufacturers are offering what is known as the Flush pipe - a standard reinforced concrete jacking pipe with an unplasticised PVC liner. Slightly less in diameter (5 mm), the liner is inserted into the pipe. The ends are closed and pressurised steam is blown in forcing the liner onto the inner wall of the pipe, attaching it to the concrete. Spigot-and-socket ends are heat-treated to ensure good adhesion.

Any sheet lining has to be designed to resist external pressures as well as internal pressures. A sound permanent bonding is required which can be provided by studs or tees on the back of the sheet, which mechanically bond the liner to the pipe. Care needs to be taken with the continuity of the liner across the joints, particularly in non-man-entry sizes.

8.3.2 Composite pipes

As some problems have been experienced with linings separating from pipes and breaking up, a number of composite forms of pipe suitable for jacking have been produced which have internal and external protective layers.

Superline pipe, described in Section 8.1.2, has inner and outer faces strengthened with a glass-reinforced cement (GRC) coating. This provides a high level of resistance to corrosion attack, which is one reason for its extensive use in Japan. Cerahume pipe, another Japanese development, is made by the centrifugal process. Ceramic material and resin are spun into the pipe to form a corrosion-resistant layer. Joints are also treated with resin.

Jacking of lined pipes is also common in Taiwan. An example is a centrifugally-made pipe lined with a mixture of crushed glass and epoxy resin, which is spun in.

8.4 Permanent access shafts

Until quite recently the working pits and the permanent access shafts were conceived as quite separate entities. At best, the contractors' pits provided holes in which the permanent access shafts could be constructed. The remainder were then backfilled. Increasingly common are drive and reception pits constructed from prefabricated concrete rings sunk as caissons, which remain in the ground to provide the basis for the permanent access chambers.

The original Japanese microtunnelling equipment typically required shafts 6 or 7 m long. For a deep sewer therefore, their construction cost represented a substantial part of the total installation cost. The demand in Europe to work from tighter locations and to improve competitiveness stimulated the design of more compact microtunnelling equipment and the manufacture of pipes in short sections. These refinements have allowed small diameter installations, such as house connections, to be carried out from shafts as small as 2000 mm in diameter. For pipes up to DN 750 it is now possible to work from shafts in sizes up to 3000 mm in diameter. Pipes up to DN 1500 mm can be driven from shafts of around DN 4500 mm.

Factory-made shaft sections are available in diameters from 2000-3000 mm, while larger sections are often cast at site. The rings are designed to be mechanically joined and hydraulically sealed. Chapter 11 discusses the construction of precast shafts. Once the installation is complete, the shaft can be furnished with a channel, benchings, access ladders and landings. The top full diameter ring can be removed and a reducing slab installed at about 1000 mm below the surface with access shaft rings and a normal manhole cover. This allows the unimpeded laying of cable and pressure pipe utilities across the main shaft of the manhole. Figure 8.21 illustrates a typical arrangement for a caisson manhole. Combining temporary and permanent shaft functions within the design can offer worthwhile cost savings.

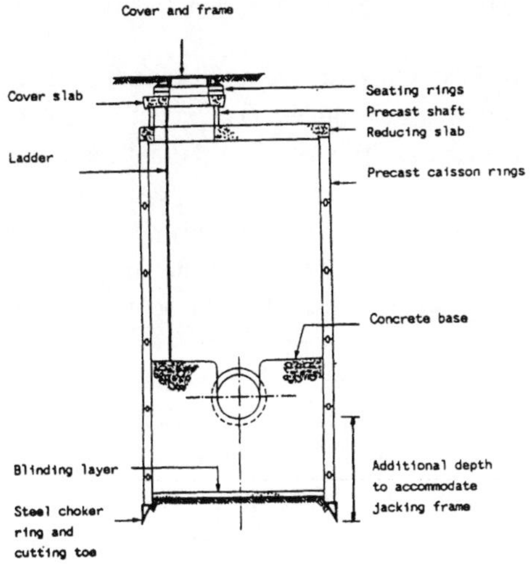

Figure 8.21 Typical precast concrete caisson manhole (after Yorkshire Water Plc)

8.5 Sewer connections

The design and layout of connections should be an integral part of a sewer scheme. It would obviously make little sense to select jacking methods to minimise disruption and then to dig a series of house connection trenches across the road. However, a more important consideration must be for the long term integrity and performance of the sewer, which to a large extent will depend on the integrity of the connections.

House connections are often the responsibility of the property owner and the work is often done independently and at a different time by local contractors and builders. Their methods of cutting into the sewer and making the connection are frequently crude. Results of studies and city records have established that lateral connections are a major source of leakage and infiltration. If making the connections were part of the integrated design, many of these shortcomings could be avoided. It is worth noting here some innovative non-disruptive techniques now being used for sewer connections. These can be divided into methods for connecting laterals directly into the sewer and methods for bringing laterals to shafts or to chambers providing access.

8.5.1 Direct connection of laterals

Connecting individual laterals from properties directly into the sewer at right angles is the traditional approach. How the individual connections for each property are made is sometimes enshrined in local and national legislation [25].

Non-man-entry sewers. For non-man-entry sewers the most common trenchless practice is to limit open excavations to a drive and a reception pit. From a small drive shaft on the property, the connection is driven to a reception pit located over the existing sewer. House connection equipment designed for this purpose is described in Chapter 5 and Figure 8.22 shows an installation of this type. The final connection into the sewer is made by hand.

Constructed from precast units of about DN 2000 mm, the drive pit can then be converted to a useful inspection chamber. The small equipment can also be taken into basements, which can be used as drive areas. Although some disruption is caused by open shafts in the road, these can be kept to minimum size and plated over when not in use.

The reception excavation at the sewer can be eliminated if the remote-control connection methods described in Chapter 5 are employed. The German company, Bohrtec, has pioneered this method [34]. However this procedure is more complex and more costly and is probably only warranted in a few cases.

Man-entry sewers. For man-entry sewers the reception shaft can usually be eliminated, as men working from inside the sewer should be able to make the final connection. Alternatively, where the diameter is sufficient to allow house-connection equipment into the sewer, a bore can be driven out to the property. However, the feasibility of this will depend on access, available space and flow conditions.

Figure 8.22 House connection equipment

8.5.2 Connection via shafts

Many authorities believe that the operational advantages of bringing connections into the access shaft are well worth the additional capital expenditure. By having access to all connections it is possible to undertake inspection, maintenance and cleaning and, when required, repair and renovation. It should be noted that some authorities and regulations ban or discourage bringing connections into access chambers.

The Berlin system. What has become known as the Berlin method [66] is a system of radiating house connections from the main sewer access shafts or, where these are not conveniently located, by constructing secondary intermediate shafts. Figure 8.23 illustrates a layout of this type, where up to four properties can be connected to each shaft. The same shaft used for microtunnelling the main sewer can be used for radially boring the house connections with the equipment described in Chapter 5.

An economic approach, and sometimes a local requirement, is that connections be kept shallow enough to enter the manhole by a *backdrop* arrangement. The integrated design has been extended to allow the space available in the microtunnelling drive shafts to be used for this. Figure 8.24 shows how a microtunnel drive shaft has been adapted to this use with backdrops positioned inside the chamber.

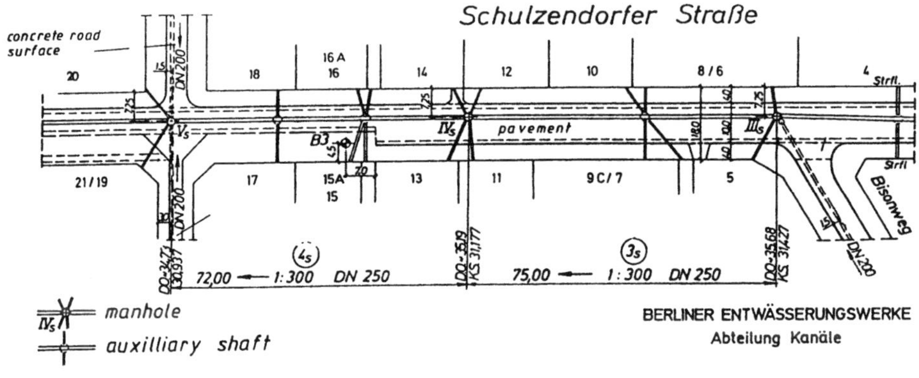

Figure 8.23 Construction of radiating house connection pipes - Berlin method layout

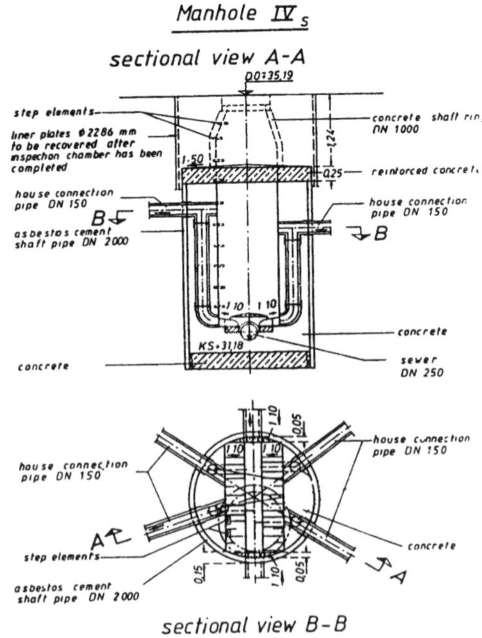

Figure 8.24 Manhole with backdrop

Collector chambers. An alternative to the Berlin method is to connect several properties by a rider sewer or radials to a collector chamber located in the side of the road. Often the radials or the rider can be quite shallow and laid by open cut without causing disruption. From the collector chamber, a single connection can be made by microtunnelling equipment to the access manhole on the main sewer. This rider sewer approach can also be used where a sewer is being replaced. The old sewer can be retained as a rider complete with existing connections and diverted into the new line at appropriate intervals as shown in Figure 8.25 [67].

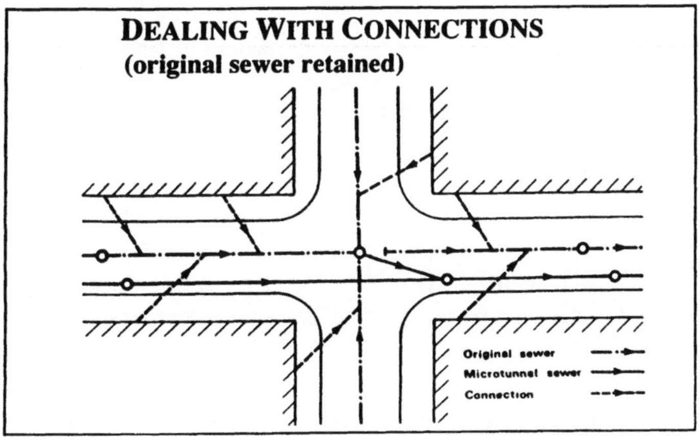

Figure 8.25 Rider sewer approach (diagram by courtesy of Yorkshire Water Plc)

8.6 Installation of a pipe within a sleeve

8.6.1 Differences in practice

A casing, or sleeve, in which the service pipe is laid is still a frequently used method of installation, particularly for crossings. The decision whether the service pipe should be laid direct in the ground or inside a casing or duct is often controlled by regulations or owner requirements. For example, US railroad companies require gas pipelines to be encased where they cross under tracks. It is perceived that the casing provides greater structural integrity and safety, although some evidence suggests that in fact the contrary may be true.

Installing a duct or casing with the service pipe laid inside requires more work and a greater cost of materials than installing a pipe direct in the ground. The additional costs are for:

- Providing and driving an external duct which must be bigger than the service required.
- Providing and installing the service pipe itself.
- Filling the annular space between the two.

The fundamental difference between flexible and rigid pipeline design was discussed earlier in this chapter. A steel casing is clearly a flexible pipe during and immediately after installation. However, after installing the service pipe and grouting up the annular space the composite structure becomes a rigid pipe. As casings are often laid at relatively shallow depths below rail tracks, highways and aircraft runways, the influence of the live loads on casings is critical and was discussed in Section 8.2.2.

The designer should always refer to regulations and design criteria which vary between countries and even from one authority to another within the same country. Some examples are given below.

USA. No less than seven bodies issue regulations pertaining to pipeline crossings under railroads and highways. The American Railway Engineering Association (AREA) and the American Society of Civil Engineers (ASCE) specifications [68] [69] recommend casings at railroad crossings. Allowable design strengths, set by the regulatory bodies, are based on the minimum yield stress and vary widely between 40% and 72%. The ASCE also recommends casings at highway crossings, but the rules of state highway agencies vary. Some allow direct installation with an increased wall thickness to the pipe. Decisions are often made by local officials about whether pipes should be cased or uncased; in practice the majority are cased. Minimum cover is another requirement which varies. Depths between 0.75 to 1.50 m have been noted.

Canada. Regulations generally follow the Canadian Standards Association (CSA) and the Canadian Transport Commission (CTC) [70] [71]. CSA Z184 for gas pipelines specifies cover based on a specified minimum yield stress (SMYS). Pipes of DN 600 mm operate under 30% SMYS and DN 1200 mm for those above. The use of casings is determined by type of location, classified into four groups according to population density and the operating percentage of SMYS.

Germany. The German ATV in association with DVGW produced ATV standard A161, which has been referred to earlier [62]. This is a highly detailed document which provides the basis for calculations for gas, water and sewer crossings under railway lines and federal highways. Calculations for direct-laid steel pipes and casings cover:

- Allowable stress
- Permissible deformation
- Buckling
- Additional stress due to out-of-roundness

The loadings for highways, railways, tramways and aircraft are all specified. The document provides a full set of tables for steel pipes, setting out the recommended wall thicknesses required under each conceivable combination of diameter, operational pressure and loading condition.

8.6.2 Filling the annular space

One of the advantages claimed for a casing is that it should be possible to contain any failures within it. Another is that a casing facilitates future replacement or repairs as the service pipe can be withdrawn. However, for the service pipe to be withdrawn later, it must be laid in a manner which ensures that it is isolated from the casing. This is achieved by providing spacer attachments around the service pipe to centre within its casing. To provide a non-corrosive environment, air and water must be excluded from the annular space by sealing the ends of the casing and filling the space with inert gas or liquid. So far, however, a permanently sealed casing has proved to be almost impossible to achieve in practice. The annular space is sometimes filled with sand or similar material, but the general practice is to grout up the annular space, which precludes any future withdrawal.

Grouting procedures must always closely follow codes of practice and suppliers' recommendations. Care must be taken to fill the annulus completely, but at the same time to avoid failure of the service pipe by implosion or movement, which can be caused by excessive pressure during grouting.

8.6.3 Cable installation

Ducts are a fundamental requirement for cable-laying. In large installations where man-entry is required, cable trays are bolted to the wall, with space allowed for maintenance. Alternatively the casing may be filled with multi-duct ways.

Concrete multiple-duct sections. Installing multi-duct sections for cable and other installations is not a new concept; two-pass methods of installation have been undertaken in the past. In the first pass, a temporary sleeve is installed by jacking or microtunnelling. In the second, permanent precast concrete sections with multi-duct ways are pushed in, displacing the temporary sections.

Tokyo Electric Power has developed this concept into a single-pass system. With an adapted earth pressure balance machine and disposing of muck by pumping, the company has installed six 150 mm cable ducts within a 710 mm pipe. During driving, the duct ways are used to carry power cables, flow and return lines and to provide a laser-beam path [72].

9 Design: choice of method

9.1 Types of failure

The designer has responsibility to ensure a safe and suitable method of installation in accordance with the job requirements. His primary concern is to avoid failure and its consequences. Other factors, such as the costs, are secondary.

Failure may occur in a number of ways but, for all underground installation, ground deformation is the chief worry for the engineer. The definition of failure changes with the location of the works. Let us suppose there are two sections of work, both involving the installation of a DN 600 mm sewer at a depth of 8 m. The soil conditions are a silty clay with the possibility of sand and silt layers. The water table is 2 m below the surface. One section is located below an open field, the other under busy rail tracks. In the first case the consequences of loss of material at the face resulting in surface settlement of 150 mm will be fairly minor. There will be no danger to life or property and compensation or remedial measures will be at small cost. The same settlement under the rail track could result in a major disaster with loss of life, with the possibility of third party damage and enormous costs of compensation. For the open field section of work, it would be in order for the engineer to opt to take the risk of choosing a method where some soil loss could occur. To do so in the second section would be professionally negligent.

Ground deformation is a function of the ground behaviour. All but a small percentage of pipejacking and microtunnelling will be located in the top 15 metres of the earth's surface and any over-excavation, excessive consolidation or collapse will be reflected at the surface in damage to pavements, structures and existing services. Face stability and the choice of shield are important factors in controlling soil movements. This chapter aims to link considerations of face stability and ground deformations.

Face stability, however, is not the only source of ground movement. The three principal sources of lost ground in trenchless pipelaying have been categorised as face and peripheral loss, consolidation and local instability.

Face and peripheral loss can occur on a continuous basis with excess material entering the shield from the face or above. Another form of continuous movement and loss can arise from ground movement into the annular space formed by the overcut.

Consolidation arises when the work is in overconsolidated clays, where long-term seepage can cause increased effective stresses, consolidation and a loss of volume.

Consolidation can also arise when working in granular material or in stone fills. The re-sorting of the material can lead to some substantial settlements without any loss.

Local instability refers to those changes that occur in the strata and lead to soil loss. Examples are encountering pockets of sand under pressure, squeezing of silty and soft clay layers and encountering artificially disturbed ground such as wells, boreholes and excavations — a fairly common event.

Although soil movement is the most important cause of failure, it is not the only one. Encountering natural obstructions (like boulders or rock outcrop) or man-made obstructions (foundations or other structures) is a relatively common cause of failure with some installation methods. Again the consequences will greatly depend on the location.

Other sources of failure include failing to install within specified tolerances or damaging the permanent pipe. Obviously such situations are of great concern to the engineer and may cause him to stop the work.

Depending on the contract arrangements and responsibilities, these types of failure and their consequences may rest with the contractor. However this can be an area of dispute. The contractor may claim *unforeseeable conditions* as the fundamental cause of the problem. The engineer has to decide if the relative cost and disruption associated with sinking temporary shafts is a justifiable solution for removing obstructions to allow the drive to continue. Such a course of action would be acceptable say under a public park but unacceptable under a major highway embankment .

It is well known in principle, but not always well accepted in construction, that return is commensurate with risk. If we want the contractor to assume the risk then we must expect to pay a premium. One alternative is to *insure* by basing assumptions on the worst possible case and finding a method, albeit at some additional cost, which will overcome the problem if it is met.

One type of failure which is entirely within the contractor's responsibility is equipment breakdown.

9.2 Theory of face stability

9.2.1 Factors influencing face stability

Face stability calculations are required to determine the likelihood of the excavated face moving or collapsing into the void created during excavation. Face stability depends upon the type and variability of the ground being excavated, the ambient stress and ground water conditions, the rate of advancement and the construction methods adopted. In order that the correct equipment and methods of construction are chosen, it is critically important that the information from the site investigation is sufficiently comprehensive to enable face stability to be predicted accurately along the full length of every drive. The methods for calculating stability of tunnel faces are generally well-established and need to be adjusted for pipejacking and microtunnelling only in terms of rate of advance and, perhaps, size of face and face support conditions.

The problem should not be treated as one of a *safe/unsafe* calculation. Excavation at the tunnel face will cause stress relief, and unless support is specifically provided to precisely match the in-situ lateral stress that has been relieved, this will result in soil *movement*. This soil movement, which can be both inward or outward, could have consequences sufficiently serious for the job to be stopped, regardless of whether the tunnel face is classified as stable. Ground displacements must therefore be considered in parallel with face stability factors. This is why the two subjects appear together in this chapter.

Face stability is calculated using a force equilibrium approach, in which the ratio of restoring to disturbing forces is calculated as a factor of safety. This should more properly be described as a factor of safety against inward movement for the majority of face stability analyses. A factor of safety of less than unity indicates instability and collapse, whereas a factor of one or more indicates stability, at least in the short term. No specific indications of movement are given by this method, with the exception that the higher the factor of safety the smaller the likely magnitude of inward movement and, as the factor rises still further, the greater the possibility of outward movement. For example if the face is supported by slurry pressure in a soil that would otherwise collapse, a low slurry pressure would result in a smaller magnitude of inward movement than would occur without support. On the other hand, a very high slurry pressure will cause outward movement and possibly blow-out. The low slurry pressure might result in a factor of safety of less than one (collapse will still occur) or marginally greater than one (collapse is prevented). The high slurry pressure, however, will result in a high factor of safety against inward movement (and a lower factor of safety against blow-out).

Another aspect of ground movement that influences face stability concerns the balance between the ground being self-supporting and the tunnel face support provided. Methods of face support during construction are discussed after the stability calculations presented below. In many soils, however, a lower face support pressure than that required to match the in-situ lateral stresses will allow inward movement of the ground, which will in turn cause arching mechanisms to develop within the soil body. This subject has already been discussed in relation to the loads acting on pipes in Chapter 8.

When tunnelling in soft ground, there are two general approaches to determining face stability: the traditional empirical approach and one based on soil mechanics principles.

The empirical ideas put forward by Terzaghi [73] are still used as a basis for determining face stability [74]. These are incorporated in a system of soil classification known as the Tunnelman's Ground Classification, reproduced in Table 9.1. Together with other geotechnical data (such as the location of the ground water table) and engineering judgement, these classifications are used to design the method of face support and construction operations.

Such empirical approaches to design are of value in understanding what is required to maintain stability. However, the soil mechanics approach is recommended as a better quantitative tool for design.

Table 9.1 Tunnelman's ground classification

Classification	Tunnel working conditions	Representative soil types
Hard	Tunnel heading may be advanced without roof support.	Very hard calcareous clay; cemented sand and gravel.
Firm	Tunnel heading can be advanced without roof support and the permanent support can be constructed before the ground will start to move.	Loess above the water table; various calcareous clays with low plasticity such as the marls of South Carolina.
Slow raveling	Chunks or flakes of material begin to drop out of the roof or the sides some time after the ground has been exposed.	Fast Raveling occurs in residual soils or in sand with clay binder below the water table. Above the water table the same soils may be Slow Raveling or even firm.
Fast raveling	In Fast Raveling ground, the process starts within a few minutes; otherwise it is referred to as Slow Raveling.	
Squeezing	Ground slowly advances into the tunnel without fracturing and without perceptible increase of water content in the ground surrounding the tunnel. (May not be noticed in the tunnel, but will cause surface subsidence.)	Soft or medium-soft clay.
Swelling	Like Squeezing Ground, moves slowly into the tunnel, but the movement is associated with a very considerable volume increase in the ground surrounding the tunnel.	Heavily precompressed clays with a plasticity index in excess of about 30; sedimentary formations containing layers of anhydrite.
Cohesive running	The removal of the lateral support on any surface rising at an angle of more than 34° to the horizontal is followed by a "run", whereby the material flows like granulated sugar until the slope angle becomes equal to about 34°. If the "run" is preceded by a brief period of raveling, the ground is called Cohesive Running.	
Running		Running occurs in clean, coarse or medium sand above the water table.
Very soft squeezing	Ground advances rapidly into the tunnel in a plastic flow.	Clay and silts with high plasticity index.
Flowing	Flowing ground moves like a viscous liquid. It can invade the tunnel not only through the roof and the sides, but also through the bottom. If the flow is not stopped, it continues until the tunnel is completely filled.	Any ground below the water table that has an effective grain size in excess of about 0.005 mm.
Bouldery	Problems incurred in advancing shield or poling; blasting or hand-mining ahead of the machine may be necessary.	Boulder glacial till; rip-rap fill; some landslide deposits; some residual soils. The matrix between boulders may be gravel, sand, silt, clay or combinations of these.

9.2.2 Stability of granular soils.

For granular soils a drained stability analysis should be carried out. The first quantitative analysis was proposed by Terzaghi and is illustrated in Chapter 8 (refer back to Figure 8.18). The major problems with this analysis concern the important simplifying assumptions about the failure mechanisms, and in particular the position of the failure planes and the stresses acting on them. Subsequent research [58] [160] has shown that very different mechanisms operate in granular soils, depending upon the relative density of the materials. These ideas are also discussed in Chapter 8.

The method of analysis recommended for calculating stability is described by Atkinson and Mair [76]. Figure 9.1 defines the parameters used in this analysis. The general equation used to calculate the required tunnel face support pressure (σ_T) for granular (cohesionless) soils above the water table is:

$$\sigma_T = \sigma_S T_S + \gamma_b D_S T_\gamma \qquad \text{- eqn 9.1}$$

where σ_S = Surface surcharge (kN/m²)
T_S = Tunnel stability number for surface surcharge
γ_b = Bulk unit weight of soil 9kN/m³)
Ds = Diameter of shield (m)
$T\gamma$ = Tunnel stability number for soil load

The tunnel stability number for soil load depends on the effective internal angle of friction of the soil (ϕ') and can be determined from the graph given in Figure 9.2. As might be expected from considerations of pipe loading, the tunnel stability number for surface surcharge depends greatly on the depth of cover (h) as well as ϕ' and can be determined from the graph presented in Figure 9.3. This graph shows that for sands and gravels, which typically have values of ϕ' greater than 30°, the value of T_s is small unless the pipe is being driven close to the ground surface. The influence of T_s can usually be ignored for pipejacking and microtunnelling operations in such soils, since the cover to shield diameter ratio (h/D_s) is typically greater than 3.0.

Research [77] has shown that the length of excavation in advance of the shield (L) has little influence on σ_T. In addition it has been indicated that the required tunnel support pressure for soil loading alone in competent granular soils is approximately independent of tunnel depth and can be estimated by considering the weight of a small triangular wedge that loosens at the crown of the tunnel. Nevertheless the equation given above provides a good basis for tunnel support pressure estimation.

If the tunnel is below the water table, equation 9.1 should theoretically be modified to

$$\sigma_T = \sigma_s T_s + [\gamma_b (h - h_w) + \gamma h_w] T_\gamma + \gamma_w h_w \qquad \text{- eqn 9.2}$$

where γ_w = unit weight of water - 9.81 kNm^3
γ' = buoyant unit weight of soil = $\gamma b - \gamma w$ (kN/m^3), and
h_w = height of water table above pipe crown (m)

However, other factors influence the forces, in this case resulting from seepage towards the tunnel face and causing collapse of the soil, in which case relatively small support pressures cannot be relied upon. The influence of water on the forces at the tunnel face is dramatic since a granular soil above the water table will commonly have sufficient water to enable suction (commonly termed *apparent cohesion*) to develop, and the material to be self-supporting. The effects on soil stability of granular soils above and below the water table have recently been researched further [78].

8.2.3 Stability of cohesive soils.

The stability of cohesive soils is determined using undrained analysis and has been traditionally carried out [79] [80] using the equation

$$N = \frac{\gamma_b (h + D_s/2) - \sigma_T}{C_u} \qquad \text{- eqn 9.3}$$

where N = Stability Ratio, sometimes termed Overload Factor, and
C_u = Undrained shear strength of soil ($kN.m^2$)

This equation assumes that the face stability is being calculated with the shield immediately adjacent to the cut face (i.e. L = 0 in Figure 9.1). It was generally concluded that the critical condition was reached when N exceeded a value of 6. This equation is still in use, although the definition of stability numbers has been refined. For example Klein [81] refers to this equation and states that pipejacking operations can be carried out without unusual difficulties if the overload factor (stability ratio) does not exceed 4-5. He also states that N values of 1-2 indicate elastic ground response in generally firm ground conditions and that soil movements in this case will be small.

For most pipejacking and microtunnelling projects, the length of open excavation in front of the face L will be minimal and the effect can be ignored. However, in cases where L has a significant value, the effect should be taken into account.

The criterion for immediate stability of a circular excavated face can be assessed using the graph presented in Figure 9.4. While this provides an immediate estimation, the use of equation 9.3 is recommended for the general case. Values of critical stability ratio (N_c), [76] or the stability ratio at collapse, have been quoted for use in the equation. These are reproduced in Figure 9.5.

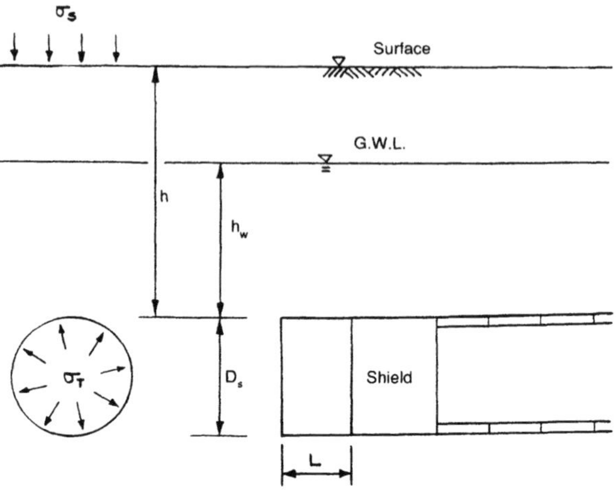

Figure 9.1 Definition of parameters for face stability calculation

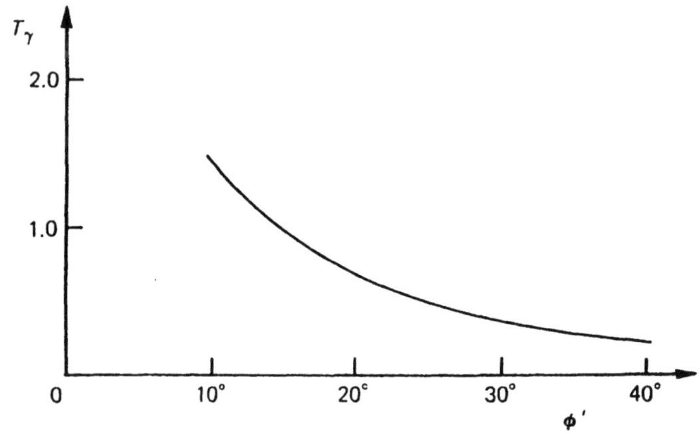

Figure 9.2 Determination of tunnel stability number in granular soils for soil load

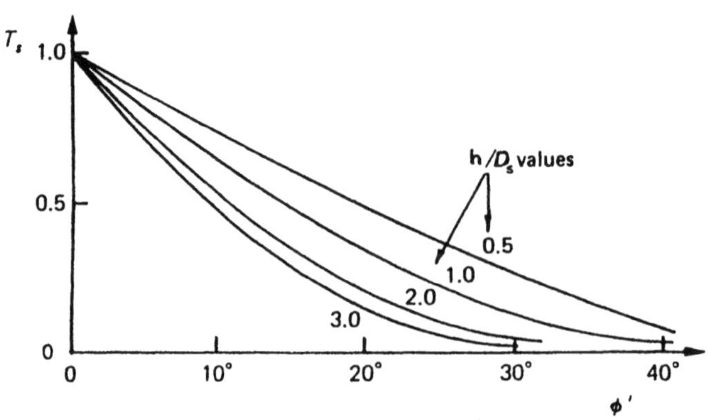

Figure 9.3 Determination of tunnel stability number in granular soils for surface surcharge

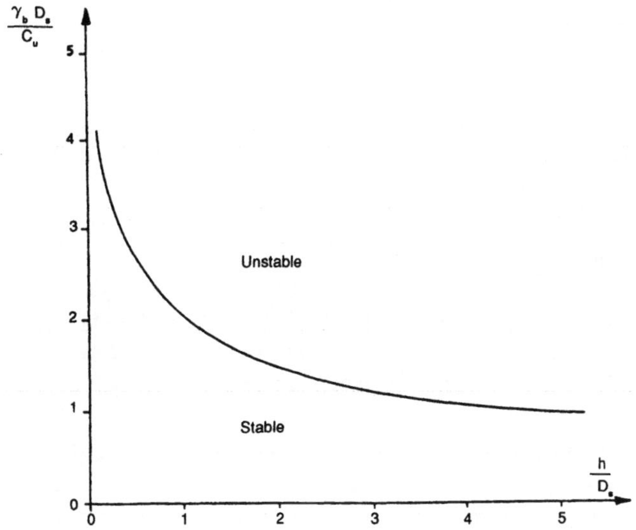

Figure 9.4 Determination of immediate undrained face stability in clay

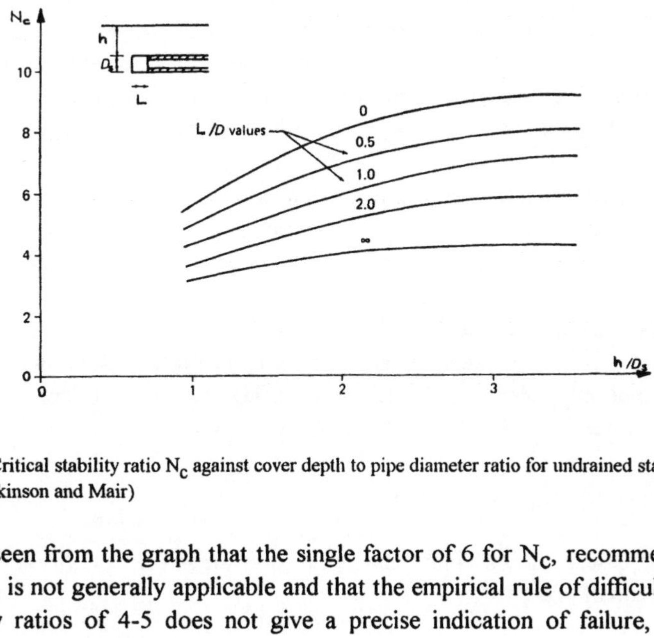

Figure 9.5 Critical stability ratio N_C against cover depth to pipe diameter ratio for undrained stability analyses in clay (after Atkinson and Mair)

It can be seen from the graph that the single factor of 6 for N_C, recommended previously for design, is not generally applicable and that the empirical rule of difficulty of excavation at stability ratios of 4-5 does not give a precise indication of failure, but perhaps an acceptable working limit. The term Load Factor (LF) is used to define stability, where

$$LF = \frac{N}{N_c} \qquad \text{- eqn 9.4}$$

Collapse occurs when LF reaches unity and thus LF as a fraction indicates the safety with respect to collapse. Clearly the value of σ_T in equation 9.3 is the only variable and this must be adjusted to provide an adequate design.

It can be seen that the curve in Figure 9.5 tends to an asymptote of $N_c \cong 9$ at cover depth to pipe diameter ratios of more than 3 and L = 0. Since most pipejacking and microtunnelling operations conform to these conditions and the shield diameter (D_s) is small in comparison with the cover depth (h), a simplified equation for collapse due to undrained instability in clay can be quoted as

$$\gamma_b \cdot h \quad = \quad 9C_u \qquad \text{- eqn 9.5}$$

giving a limiting value of undrained shear strength (C_u) to maintain stability. If a lower value (say N = 4-5) is used, this in effect is utilising a factor of safety of 2 on the undrained shear strength and the design is inherently tending to reduce likely soil movements.

Soft, weak alluvial or organic clays, which typically have low undrained shear strengths, will apparently collapse more readily than heavily overconsolidated clays or tills which typically have very high undrained shear strengths. When deciding upon a suitable strength parameter for design, the nature of the cohesive soil (in particular the soil structure and permeability) should be assessed since undrained analysis refers to the immediate condition only. If a cohesive soil has a high permeability due to features within its structure (e.g. sand or silt lenses) or to its inherent material properties (e.g. clayey silt or clayey sand) then the undrained strength measured by rapid laboratory or in-situ tests will reduce with time after excavation and the stability number calculated will rise. The rapidity with which N rises will determine the length of time that the excavation remains stable.

Stability can be maintained by applying to the tunnel face a support pressure that is less than or equal to the lateral stresses acting in the ground. However this analysis assumes that the at rest lateral earth pressure is equal to the vertical stress (i.e. the coefficient of lateral earth pressure at rest, K_o, is equal to unity). If the pipeline is being constructed in a heavily overconsolidated clay soil, K_o will rise above one and could approach a value of 2.5 in severe cases. This is unlikely to cause problems in practice since these materials have a relatively high inherent strength and the restoring force will be large. The force equilibrium equality quoted as equation 9.3 can be rewritten as

$$\delta_T \quad = \quad \gamma_b (h + D_s/2) - N_c C_u \qquad \text{- eqn 9.6}$$

There is a second case in which failure can occur and this is the case of blow-out, for which the critical equation is

$$\delta_T \quad = \quad \gamma_b (h + D_s/2) + N_c C_u \qquad \text{- eqn 9.7}$$

Mair [82] [83] has considered this case by producing curves of δ_T/C_u against h/D_s for different ratios of $\gamma_b D_s/C_u$, an example of which is shown in Figure 9.6. Mair recommends a factor of safety of 1.5 to 2.0 to ensure control of ground strains, and ultimately surface movements. The limits for acceptable heave/settlement assuming a factor of safety of 2.0 are shown in the figure. Ground movements will not occur if the in-situ stress is matched by the tunnel pressure, and this case ($\delta_T = \delta_v$) is also shown on the figure. As the consequences for surface heave/settlement are greatest at shallow depths, it will be necessary to control pressures, in this case within a narrow range of safe values.

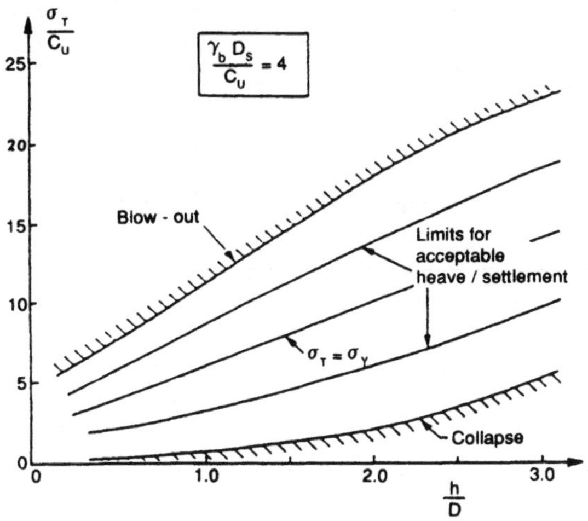

Figure 9.6 Graph for calculating the face pressure required for stability in clay.

9.3 Ground deformation

Many of the considerations concerning ground deformation have been discussed at length in Sections 8.2 and 9.2.1. This section aims to summarise the issues before going on to describe methods of estimating the movements occurring in practice. Ground movements originate from the two primary sources of (a) volume loss (or gain) at the face and (b) settlement of soil into the overbreak void after the shield has passed.

Construction support pressures (slurry and earth pressure balance) and ground treatment as ways of controlling the ground are discussed in Section 9.4. The extent of the ground movements associated with volume loss at the face will be determined by the tendency of the soil to dilate or contract when it is subjected to shear forces. In a highly dilatant dense angular sand with relatively little ground loss at the face, for example, the ground movements will attenuate rapidly and will reduce to zero within approximately two pipe diameters of the shield. In a very loose (contracting) sand the movement, particularly upwards, will extend over a much larger distance and will probably reach the surface of the soil. The reason for this latter behaviour is that the disturbance caused by excavation allows densification of the soil and a volume of ground loss at the surface greater than that at the face [84] [85]. Where movement is away from the face due to overcompensation of fluid and/or earth pressures, the converse is true. In this case the movements will extend much farther in dilatant soils and will attenuate rapidly in soils that readily contract when sheared and compressed.

In granular soils such behaviour occurs almost immediately. In clays, undrained shearing will tend to occur, resulting in constant volume deformation transmitted to the surface

[85]. Long-term consolidation or dilation can also take place depending on the magnitude and type of pore water pressures generated. A good example of this was demonstrated by a pipejacking operation in very soft alluvium [161] in which overcompensation of the in-situ stresses resulted in surface heave during the construction operation (Figure 9.7). The high positive pore water pressures generated during the construction process resulted in subsequent consolidation settlement.

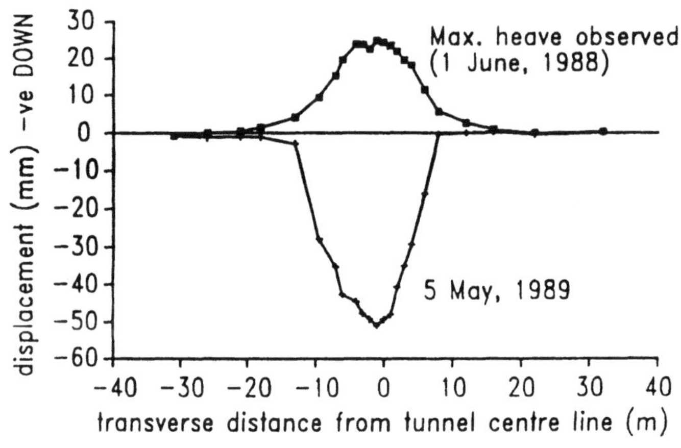

Figure 9.7 Surface heave and subsequent settlement when pipejacking a 2.1 m diameter tunnel at a depth of 10 m in very soft alluvium at Tilbury, London (after DeMoor and Taylor (1991)

Settlements occurring as a result of the overcut are similarly controlled by the tendency of the soil to dilate or contract when sheared. Dilation results in a small affected zone concentrated in the area above the pipeline, whereas contraction is associated with movements extending, essentially vertically, up towards the surface of the soil [75] [160]. Rogers and Chapman also show that the soil tends to be drawn along the overcut void as the pipe is advanced, due to friction between the pipe and the soil, thus reducing the facility for vertical settlement. These comments apply most to behaviour in granular soil. In cohesive soils undrained, constant volume shearing would be expected, although the tendency to dilation in very stiff, fissured clays might be revealed immediately after excavation. Settlement into the void can be reduced by the injection of lubricants into the overcut void as the shield moves forward. Settlements caused by the overcut are usually small and can be reduced to negligible values by filling the void. The ground movements caused by properly controlled pipejacking and microtunnelling operations are, in any case, relatively small in comparison with those caused by trenching [86].

One simple means of illustrating the ground movements caused by face pressure is shown in Figure 9.8, although this representation is perhaps most indicative of the undrained, constant volume shearing that is likely to occur, for example, in a soft alluvial clay. The figure shows that a face support pressure lower than the active pressure of the soil will result in significant ground loss which is transmitted to the surface as a settlement trough. Conversely if the face support pressure is greater than the passive pressure of the soil, outward movement at the face will result in heave at the surface. The balanced situation is thus achieved where the face pressure lies between the active and passive case. This rather simple model serves to introduce the methods of predicting surface settlement and surface heave, and, via recent research, the prediction of subsurface ground movements.

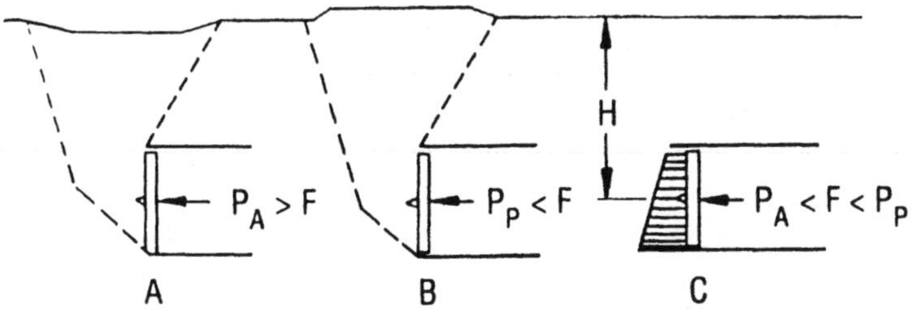

Figure 9.8 Diagrammatic representation of ground movements caused during construction operations.

Predicting surface settlement requires the practical, engineering judgement that comes from a good knowledge of the ground conditions and the construction methods to be employed. Two fundamental approaches are (a) the empirical methods based on normal probability, or gaussian distribution, curves [87] [88], and (b) theoretical methods based on fluid flow [89] and adapted for trenchless pipelaying [84] [86].

With both of these approaches the volume loss at the face needs to be estimated before calculations are made. Attewell and his researchers [87] present a table of 68 case histories on which to base judgement and O'Reilly and New [88], having analysed case history data, present a simplified method for predicting ground loss. The ground loss at the overcut is simpler to predict since it is necessarily limited. The overcut volume should be used to give an upper bound value.

The normal probability distribution curve for predicting surface settlement is shown in Figure 9.9 in which the maximum surface settlement (δ_{max}) occurs above the tunnel centreline. The geometry of the curve is defined by the maximum surface settlement and the point of inflexion (i), or maximum gradient, which is assumed to be equivalent to the *standard deviation* of a normal probability distribution. The width of the surface settlement trough, which necessarily reaches a smooth asymptote at the original ground level, is commonly assumed to occur at a distance of 2.5 i from the centreline. The volume of the settlement trough (V_s) is approximated by

$$V_s = 2.5\, i\, \delta_{max} \qquad \text{- eqn 9.8}$$

For undrained deformation, V_s will be equal to the volume loss at the tunnel face (V_t) and thus

$$V_t = 2.5\, i\, \delta_{max} \qquad \text{- eqn 9.9}$$

From normal probability theory, the settlement at any point on the curve (δ) at a distance x from the centreline can be found using

$$\delta = \delta_{max} \exp\left(\frac{-x^2}{2i^2}\right) \qquad \text{- eqn 9.10}$$

It is apparent, therefore, that both the maximum settlement and the settlement profile can be determined if the volume of the settlement trough, and hence volume loss at the face, can be estimated.

Estimating face loss remains an empirical process and is affected by the many construction and ground factors discussed earlier. Designers tend to err excessively on the side of caution. A better approach would be to produce both *best guess* and *worst case* estimates and refine the prediction by careful engineering interpretation. The value of V_t is typically calculated as a percentage of the volume of excavated material, and thus an appropriate percentage has to be estimated for the relevant ground and excavation/support conditions. As there is little available information on pipejacking and microtunnelling projects, advice on appropriate values for design will have to be taken from literature on tunnelling.

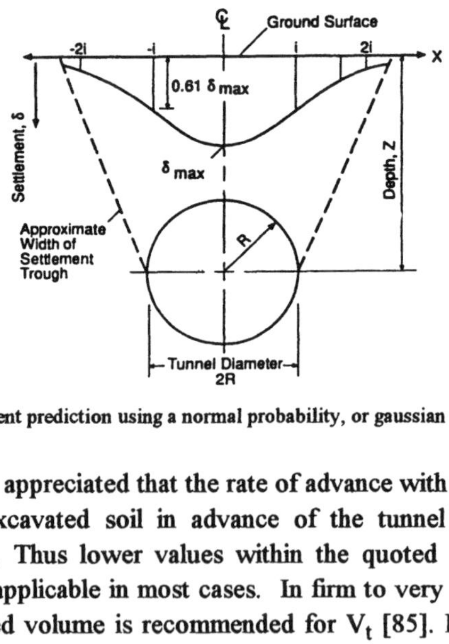

Figure 9.9 Surface settlement prediction using a normal probability, or gaussian distribution, curve.

However it should be appreciated that the rate of advance with these techniques is greater, and the length of excavated soil in advance of the tunnel is typically far less, than traditional tunnelling. Thus lower values within the quoted ranges of volume loss for tunnelling should be applicable in most cases. In firm to very stiff cohesive soils 0.5% to 2.5% of the excavated volume is recommended for V_t [85]. Klein [81] suggests that V_t should be predicted on the basis of the stability ratio (N) and that $V_t \cong 1.0\%$ where N is less than 4. For N values of 4 to 6 Klein suggests that V_t will lie between 1% and 5%. A more precise approach, offered by Attewell and Yeates [85] is reproduced in Figure 9.10. Glossop's (1977) expression for volume loss is

$$V_t(\%) = 1.33\,N - 1.4 \qquad \text{- eqn 9.11}$$

for N = 1.5 - 4.0 [90]. For cohesive soils, undrained shearing, and hence equality between V_t and Vs. can be assumed for design. Equation 9.9 can therefore be used directly.

In non-cohesive soils, the choice of volume loss is much more uncertain. Attewell and Yeates suggest a range of 3% to 10%, with a typical value of 5%, whereas Klein suggests a value of 1%, provided that heading stability is maintained and ground water conditions are positively controlled. In the operations considered here, Klein's value would appear the most relevant for design. In non-cohesive soils and at typical cover depths to pipe diameter ratios, however, the inherent dilation (typically) or contraction (occasionally) of the soil will result in very different volume losses at the surface. In dilatant soil much, and sometimes all, of the volume loss at the face will be taken up by soil dilation and no effect will be registered at the surface. In loose non-cohesive soils, contraction on disturbance could cause the surface loss to be significantly greater than that at the face, and could exceed 10%. Greater engineering judgement should thus be used in loose soils. An even higher range for volume loss is recommended for man-made ground and other fill materials.

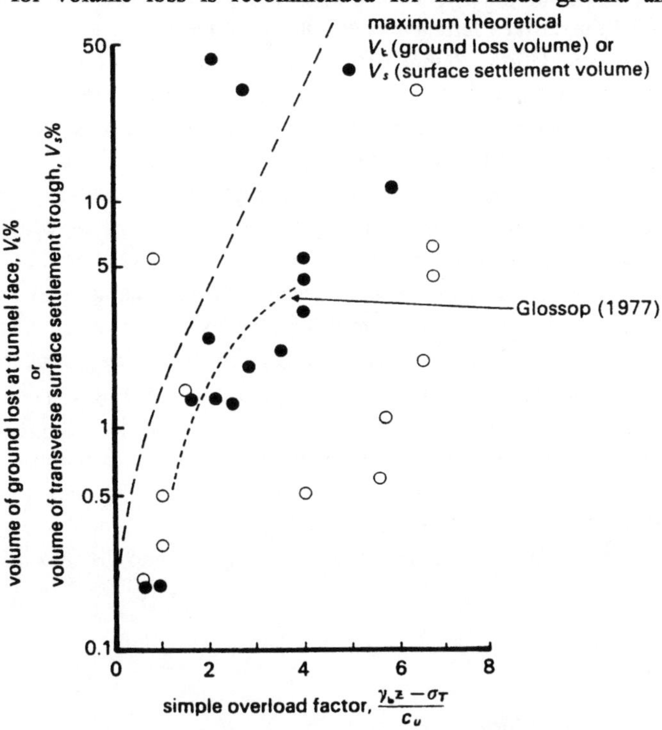

Figure 9.10 Prediction of ground loss based on the stability ratio for cohesive soils (after Attewell and Yeates, 1984).

The settlement trough width (2 i) similarly requires estimation for design, and this depends upon both the depth of cover (h) to shield diameter (D_s) ratio and the shear strength of the soil near to the surface. O'Reilly and New [88] recommend the expression

$$i = K z \qquad \text{- eqn 9.12}$$

where K = constant

z = depth to effective source of ground loss (m), usually taken as $h + D_S/2$.

Values of K of approximately 0.5 are recommended for cohesive soils, with typical values of approximately 0.4 for stiff clays ranging to 0.7 for very soft clays, and 0.25 for non-cohesive soils. The study of site measurements yielded the more precise expressions of

$$i = 0.43\,z + 1.1 \qquad \text{- eqn 9.13}$$

for cohesive soils with very strong correlation, and

$$i = 0.28\,z - 0.1 \qquad \text{- eqn 9.14}$$

or granular soils with a lesser, though still acceptably high, degree of correlation.

Surface settlements based on estimated values of V_s and i can easily be assessed using equation 9.8. To the surface settlements calculated in this way, should be added the long-term effects of consolidation. These effects result chiefly from dissipation of pore water pressures generated during excavation and construction. However long-term effects can also result from the new pipeline acting as a drain to the surrounding soil or, (more likely for the methods considered here) the disturbed, typically loosened and more permeable soil acting as a drain. The ground movements at depth, which will affect buried services like adjacent pipelines, can be, approximately rather than accurately, estimated from the surface settlement profile by simple proportioning with depth [84].

The second approach [86] uses fluid flow techniques for radial ground movements, and incorporates directional (predominantly vertically downward) movements, and contraction and dilation effects [84]. Once the volume loss is estimated, its distribution should be defined. This may be uniform radial, uniform radial over the upper half of the pipe or variable from a maximum at the shield crown to zero at the shield springings. The flow of the ground into this void can then be calculated using relatively simple equations. This technique has produced good correlations with field data. The surface profile, being similar to that produced by the normal probability method, is thought to better estimate the sub-surface movements, particularly those occurring close to the shield.

9.4 Control of face stability

Three approaches can be taken to maintaining face stability for pipejacking and microtunnelling:

(i) To apply a ground treatment technique prior to construction.
(ii) To choose a shield or TBM which provides the necessary support and balance against soil and ground water pressures.
(iii) To combine (i) and (ii).

9.4.1 Ground treatment

In designing a pipejacking or microtunnelling project, the possibility always exists for prior treatment of the ground to provide stable working conditions. Pre-treatment of the ground is common in open-cut and shaft sinking but less common in pipejacking and microtunnelling, because to a great degree these methods can provide their own inherent stabilisation. However, in some ground conditions, pre-treatment may be necessary in combination with the selection of a pressure balance system. Some form of pre-treatment may prove to be a cost-effective way of stabilising the ground before excavation commences.

Dewatering. Where there is a high water table, the level can be temporarily lowered to about or below invert in the area of the installation. This can be done in granular soils such as gravels, sands and cohesionless silts with a high permeability.

Vacuum wellpoint dewatering is commonly used as a means of temporarily lowering ground water but it has some disadvantages. First, it is limited to a working depth of 6 m. Second, as the dewatering pipe occupies a considerable surface area over the line of drive, the non-disruptive advantages of pipejacking or microtunnelling can be offset. Other inhibitors are its cost of installation and the fact that it needs to be operated on a 24-hour basis.

An alternative is deep well pumping with submersibles, but the applications are more limited. The wells must be installed in advance of the work, as lowering of the water table can be a lengthy process.

The major drawback of all forms of dewatering is that reducing the water table can cause ground consolidation and surface settlement. There are also increasing environmental objections to its use.

Grout injection. Materials either in suspension or in solution can be injected into the soil to decrease its permeability, to increase its shear strength or to reduce its compressibility. Grout injection by various means is used in underground works by the civil engineering and mining industries.

Permeation grouting requires drilling down to the target area and pumping into the soil a liquid solution which displaces the pore fluids and then solidifies. Pore size is the key engineering factor in this kind of permeation grouting as it will determine the type of grout to be used as well as the spacing of injections and the appropriate pressures. Local variations in pore size and permeability may mean that some areas are left untreated.

For more open permeable soils, materials such as cement, bentonites and PFA's can be injected economically. However, with decreasing permeability, more expensive materials have to be employed, including silicates, phenolic resins, lignins, Acrylamides and Aminoplasts.

A number of pipejacking works have been driven in ground pre-stabilised by grouting. Most often the ground is treated in this way at the break-out area immediately outside the shaft.

Freezing. The freezing of soil to make it temporarily more stable is more common in major tunnelling or shaft-sinking works, but the scale and cost of such an operation do not generally lend themselves to trenchless installation works. However, there are case histories of pipejacking installations where ground freezing has been used to stabilise the ground [91]. Figure 9.11 summarises ground treatment techniques and the range of soils to which they can be applied. For more extensive information on ground treatment reference should be made to the books by Tomlinson [92] and Koerner [93].

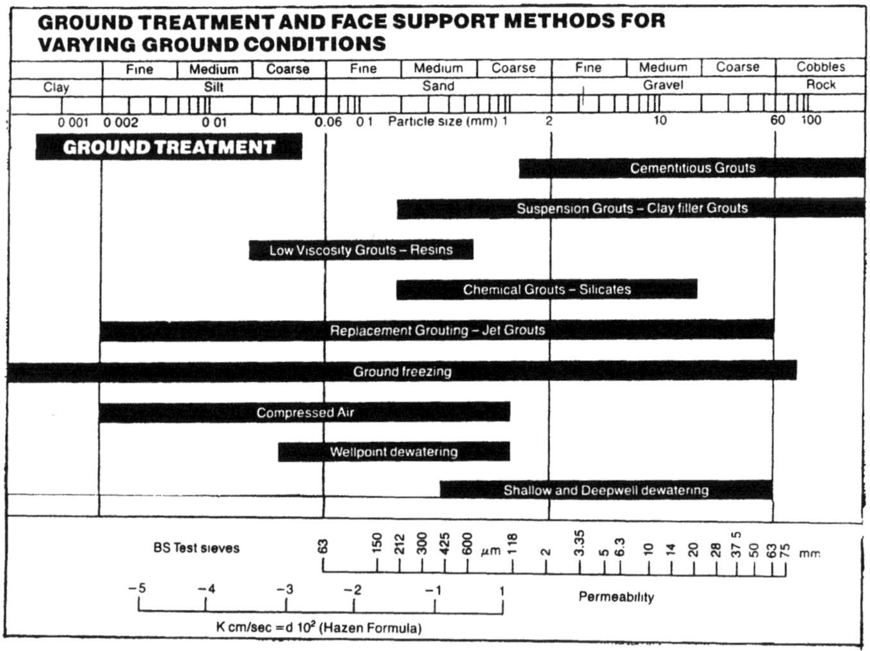

Figure 9.11 Ground treatment methods for varying ground conditions (after PJA Manual).

9.4.2 Choosing a shield or TBM

The range and diversity of pipejacking and microtunnelling machines has already been indicated in Chapter 3 and refinements continue to be introduced every year. The importance of site investigation and soil analysis has been discussed at both theoretical and practical levels.

To select the most effective type of shield and TBM for a specific installation in the ground conditions predicted, four primary considerations will be:

- The stability of the face.
- The capability to deal with changing ground conditions.
- The ability to deal with obstructions.
- The overcut.

The contractor will be concerned with the technical features offered by the alternative shields and TBMs available to him. Chief among these will be the relative merits in their steerability, soil cutting efficiency, soil removal efficiency and their rate of progress In this section an attempt is made to offer some guidance on the suitability of the main types of shields for working in certain types of soil conditions. For this purpose, shields are divided into three categories:

(i) Pipejacking: man-entry shields with access to the face.
(ii) Pipejacking: pressure balance shields without access to the face.
(iii) Microtunnelling: non-man-entry.

Pipejacking: man-entry and access to face. The influence of varying conditions and the impact on stability numbers is discussed in Section 9.2. Stability numbers can be a useful guide when working with open shields in cohesive soils. Put simply, in homogenous clay an open face will fail absolutely when the weight of the overburden at the crown reaches nine times the undrained strength of the clay. To provide a safety factor (minimum 1.8) stability numbers of 4 to 5 are suggested in soils with a high undrained shear strength, such as laminated clays and glacial till. Materials like soft weak organic or alluvial clays, with low undrained shear strengths, become more of a problem.

When using a non-pressurised open shield in cohesive soils it may be found that for a stability number less than 5, there may be face instability where rates of advance are low or the soil is of low plasticity. This is due to drainage of the face soil which in the case of clay causes the strength to change from cohesive to frictional. The time taken for the face to drain will depend on the permeability and compressibility of the soil. When permeability is high, drainage is rapid, and stand-up time will be short. Conversely in low permeability soils stand-up time will be longer. To avoid these time-dependent face stability problems in cohesive soils, the rate of machine advance should be high enough to ensure that undrained conditions prevail. As the cohesive soil becomes less plastic, and therefore more permeable, rates of advance have to be increased to maintain undrained conditions. Where this is not possible then some form of pressurised support of the face is necessary to avoid instability.

In practice, open-face jacking with various types of shield configuration is applicable to a wide range of soil conditions. This type of shield has the major advantage of being able to cope with most forms of obstruction, as it is possible for the miners to identify them and then to remove them whole or break them up. Open shields sometimes have doors as a preventive device against emergency rather than as a face control mechanism.

Other features for face control include spiling, compartments, breast plates and shelves. However, their use in installations of 1500 mm diameter or less is limited by lack of space.

Blind shields, although classified as open shields, often have no access, or at best limited access, to the face. Blind shields are only suited to working in highly plastic soil with N values, say, between 0 to 3. Controlling the volume of the remoulded soil as it is squeezed through a face aperture would appear to involve complex judgements. Such shields are not recommended in situations where ground movements are critical.

Semi-mechanical shields with a boom cutter or hydraulic back hoe represent another form of open shield. These increase drive and therefore production rates, which is an obvious economic advantage. The increased rate of driving can be technically beneficial in some conditions.

Mechanical rotary shields are full-face wheels, discs and domes. They are basically open-shield machines, but with limited access to the face, which makes it more difficult to deal with obstructions. In some forms they are almost remote-control with disc and dome heads providing some face support. These machines are at their best in homogeneous stable material and can achieve high rates of driving. They are not suitable for work below the water table and in soft material they have a tendency to nose down and be difficult to steer.

Compressed air is finding increasing use in jacking as it offers a combination of pressure balance with access to the face. Such a combination can be valuable where obstructions are known to exist in unstable soil with a low permeability. It is an advantage to have a system where miners can work in free air while using remotely controlled equipment at the pressurised face. Provision can be made for man-access to the face if boulders or other obstructions are encountered.

The air lock is located in the line and travels forward with the jacking. However, the relatively low volume of air within the locked section makes air pressure working unsuitable for shallow installations in permeable ground or where voids are likely, as pressures can fluctuate sharply. Work is usually only carried out with very limited pressures (1 bar or less). Compressed-air working should not be undertaken without expert advice and fully taking into account national and local regulations. Further information on working in compressed air is given by Nussbaumer [94].

Pipejacking: pressure-balance shields without access to the face. Pressure shields balance the pore water and soil pressure by providing a chamber where water, mud, soil, air or mechanical thrust provides the counterbalance. There is no access to the face.

Unwanted consolidation can occur through several mechanisms including loss of ground water into the tunnel face where face support fluids such as bentonite slurry do not precisely balance ground water pressures. Permanent surface settlement can be caused where face instability is allowed to develop. Indeed unless the earth pressure at the face is precisely balanced there will always be movement at the ground surface. If the face is allowed to move towards the shield when a tunnelling shield is advancing, it will result in a zone of surface settlement travelling in front of the shield. Conversely if the shield is forced forward into the face so that soil moves away from the shield, it can result in a zone of surface heave travelling in front of the shield. The required balance between these two extremes is a shield which advances at such a rate that the soil neither moves towards the shield nor is thrust away. The pressure at the shield face should ideally be maintained at the at rest pressure to avoid horizontal compression or expansion of the soil and prevent settlement or heave. See Section 9.3 for a detailed discussion of the mechanics of ground movement.

Of the increasing number of pressure balance machine designs, some are directed towards a relatively narrow band of geotechnical conditions and others to a broader range.

Slurry chamber TBMs employ both water and mud to form the pressure balance slurry. For many cohesive soils like clays and silts, water mixes with the material to provide an adequate slurry. Water will be a suitable fluid when working below the water table. Above the water table in non-cohesive soils, as particle size and permeability increase, it will become necessary to use designed muds which will build up a filter cake at the face and not be dispersed. For open-structured gravels and boulders, preliminary stabilisation grouting may be required. Some slurry machines have not performed well in firmer clays.

Machines like the Unclemole are designed to work in sandy soils, sands, gravels and cobbles with N values in the range 5 to 50. Various head configurations, slurries and crushing devices are available to work in soils ranging from clays through to coarse granular materials and rock. Some machines incorporate crusher devices which can deal with cobbles and individual larger boulders up to 30% of the shield diameter. Ground water heads of up to 30 m (and more with special arrangements) can be safely worked.

It should be pointed out that slurry pressure balance machines are now often selected not just for the geotechnical advantages of pressure balance in difficult soil conditions but also for their technical and economic efficiency in stable soils.

Earth pressure balance shields are most suited to soft and unstable formations including high plasticity clays, silts and granular soils with a high percentage of fines. Machines like the Iseki MEPCB, which combines a slurry chamber with mechanical earth pressure balance, are frequently employed where ground movement is a primary concern. These machines are less well suited to highly permeable soils and where there are hard inclusions.

Microtunnellers: non-man-entry. The three main types of microtunnelling equipment described in Chapter 5 (pilot, auger and pressure balance) are each broadly suited to differing ground conditions

Pilot-bore microtunnellers have probably not been adopted greatly outside Japan because they are mainly suited to homogeneous soft ground conditions with relatively low N values. In general they are used for smaller installations (below 250 mm).

Auger excavation systems are also chiefly confined to smaller microtunnellers, although a few larger versions will install sizes over 900 mm. In their simpler forms they are equivalent to an open shield and for the most part perform well in stable cohesive soils. Without incorporating special measures they are not suited to working below the water table with medium to high permeability soils or in soils with N values below 5.

However, more sophisticated auger machines are being provided with capabilities of temporary face pressurisation using compressed air and fluids. Other machines have the auger head and flights designed to operate under a water balancing fluid pressure. Another means of pressurisation is to incorporate lead augers with a closer pitch which gives an earth pressure balance effect at the head. Auger type machines which have a chamber at the face, with a separate drive head and slurry pumping, appear to offer many of the advantages of the pressure balance machines. Auger microtunnellers should not normally be used where there is more than 3 m of ground water head.

Slurry and earth pressure balance microtunnellers are essentially smaller versions of the pressure balance pipejacking TBMs referred to in this section. Most of the comments already made apply also to the microtunnellers, although the particle sizes that can effectively be handled will be more limited. Many slurry machines have the capability of dealing with cobbles and other obstructions up to 30% of the shield diameter as was the case with the Unclemole. Working has also been extended into much harder materials including solid rock by the development of rock cutting heads. Table 9.2 summarises the ground conditions suitable for small diameter pipejacking and microtunnelling equipment.

9.5 Rock

Working in rock is less complicated than dealing with occasional boulders, particularly since machines and heads have been designed specifically to work in material classified as strong to very strong. Some standard machines can also be fitted with alternative heads for driving through rock.

To work in rock, as opposed to occasional hard conditions, requires a machine to be purpose-designed with a variable speed, independently moving head, and stabilisers. Not unexpectedly, it is the wear on bits in hard and abrasive conditions that can limit the drive. To overcome this, some machines have provision for replacing cutters from the back of the head [95]. Considerable progress has been made with developing roller and disc bits to give longer life and greater capability in the harder rock. The Iseki Rockmole, designed to work in diameters of 600 mm and greater, is claimed to have a capability to drive full face in rock strengths up to $100 N/mm^2$ and for short periods in rock with strengths up to 200 N/mm^2. Auger-based machines using button bits are normally used for much lower rock strengths.

Pipejacking in rock, though economically and technically viable, requires the correct method of installation to suit the characteristics of the rock concerned. Particular attention therefore needs to be paid to the Rock Quality Designation (RQD), which is a measure of the degree of fracturing. The choice of equipment and its bits (roller, disc, button etc.) will be based on an assessment of the basic rock type, its RQD and its compressive strength.

Table 9.2 Matrix of microtunneller types and ground conditions.

GEOTECHNICAL			METHOD			
Classification	Description	N value	Pilot	Auger Standard	Auger Independent drive	Slurry
High plasticity clays	Peat/organuc	0	◐	●	◐	◐
	Very soft	0-2	○	◐	◐	○
	Soft	2-5	○	○	○	○
	Firm	5-10	○	○	○	○
Medium to low plasticity clays	Firm to stiff	10-20	◐	○	○	○
	Stiff to hard	>20	●	○	○	◐
Fine granular	Very loose	<4	●	●	◐[2]	○
	Loose	<10	◐	◐[1]	○[2]	○
	Medium dense	10-30	◐	○[1]	○[2]	○
	Dense	>30	●	◐[1]	○[2]	○
Coarse granular	Medium dense	10-30	●	◐[1,3]	○[2,3]	○[3]
	Dense	>30	●	●	◐[2,3]	○[3]
Rock	Weak	>100	●	●	◐[4]	◐[4]
	Moderately weak	>200 (>5MN /Nmm	●	●	◐[4]	◐[4]
	Moderately strong	>15MN /Nmm	●	●	◐[4]	◐[4]

Legend

○ Suitable

◐ Possible

● Not suitable

Notes

1) Standard auger not suitable for work in granular soils below water table.

2) Below the water table, augers with pressure balance devices have some limited capability.

3) All machines have max particle size limitations in coarse granular soils.

4) Modified for heavy duty working with rock head.

10 Temporary work design: the line

10.1 Jacking loads

The jacking load is the force required to advance the complete pipe train after, or during, excavation of soil at the tunnel face. The load is a combination of face resistance to penetration of the shield, and frictional force on the following pipeline. It is important that jacking forces can be accurately predicted in practice to enable the operation to be designed appropriately.

Jacking and reception shafts are located according to hydraulic considerations, the positioning of manholes, the requirements of maintenance and restrictions imposed by the pipejacking or microtunnelling construction techniques. These latter restrictions include the requirements of curved drives where jacking loads are typically more concentrated, the progressive loss of production in longer drives, and limits on length imposed for recovering ill or injured operatives at the face. Once the spacing between the shafts, and thus the length of the drives, is known, an estimate should be made of the jacking forces, from which the pipes, the jacking equipment and the thrust wall can be designed. The load-carrying capacity of the pipes, under the worst likely degree of misalignment foreseen, must be sufficient to withstand the forces predicted. This is discussed further in section 10.1.3. Practical measures to avoid overstressing the pipes include injecting lubrication and providing interjacking stations. Failure to predict the jacking forces accurately can result in overstressing of the pipes and/or lack of thrust capacity in the rams, as well as failure of the thrust wall.

Predicting these forces is made more difficult by the uncertainty of ground conditions that besets every geotechnical engineering design. Uncertainty is more likely in tunnelling as it concerns an underground linear structure, therefore some overdesign is necessary to allow for this. However this is no excuse for the engineer to evade his design responsibilities by grossly oversimplifying the prediction of jacking forces. The engineer must be concerned with the economy of the construction operation, and thus the cost of the finished structure to the client and/or profit to the contractor.

The engineer has to analyse (a) the forces that arise from the prevailing site conditions and (b) how these forces are influenced by the construction operations.

Forces determined by design or site conditions include:

- The size, shape, self-weight and external surface of the pipes.
- The length of the pipeline to be installed.
- The type of soil and its variation over the length of the drive.
- The position of the water table.
- The stability of the soil, both immediately and over the construction period.
- The cover depth and unit weight(s) of the overlying strata.
- Surface surcharge and vibrationary loading.

Construction-related factors that influence the magnitude of the jacking forces include:

- The amount overcut during excavation.
- The use of lubrication.
- Steps at joints and/or joint deformation.
- Jacking around curves.
- Misalignment of the pipeline along its length.
- The use of intermediate jacking stations.
- The rate of advancement of the pipeline.
- The frequency and duration of stoppages.

Most of these factors influence only the frictional resistance along the pipeline, while a few also influence the face resistance to penetration of the cutting head and shield. These factors will be considered in more detail in the following two sub-sections.

10.1.1 Face resistance

The face resistance derives from different sources in open-face and pressure balance operations, and these cases will be dealt with separately below. Open-face operations include manual excavation and cases where auger or cutter heads are used. Pressure balance operations include those that employ earth pressure balance, earth pressure plus slurry pressure, or slurry pressure alone.

Open face operations. In open face operations, the face resistance is solely due to the penetration resistance of the cutting edge and the friction acting on the external surface of the shield. In a cohesive soil the excavation is typically limited to slightly within the external diameter of the shield. As the shield advances, it trims the excavation to precise dimensions. In a non-cohesive or unstable soil the shield is thrust forward with a constant force into the soil to ensure that the face does not collapse. In a cemented soil or rock, excavation is carried out ahead of and to a larger diameter than the shield, in which case the face resistance is relatively small as it is due solely to the frictional forces acting on the external surface of the shield.

Cutting edge resistance has been researched extensively in Germany. A method attributed to Herzog [96] is based on the observations of driven piles. These observations were translated from the vertical to the horizontal plane via successful correlation with jacking site records. The cutting edge resistance (P_S in kN) is given simply as the product of the cross-sectional area of the cutting edge and the tip resistance (p_S), which is tabulated for different soils in Table 10.1. The equation is

$$P_S = \pi \cdot D_S \cdot t_S \cdot p_S \qquad \text{- eqn 10.1}$$

where D_S = External diameter of shield (m)
t_S = Thickness of cutting edge of shield (m)

Table 10.1 Tip resistance for calculation of cutting edge resistance (after Herzog, (1985)

Soil type	P_S(kN/m²)
Soft rock, cemented soil	12000
Gravel	7000
Dense sand	6000
Medium sand	4000
Loose sand	2000
Stiff to hard clay	3000
Soft to firm clay	1000
Silt, alluvium	400

The reason for the high value for rock given in the Table is that Herzog is considering penetration into the rock, rather than sliding into a void created in the rock. A nominal value for face resistance when pushing into a void can be calculated from

$$P_S = \frac{W_s \tan \delta_s}{\cos \zeta} \qquad \text{- eqn 10.2}$$

where W_S = Weight of the shield (kN)
δ_S = Coefficient of friction between steel and rock (degrees)
ζ = Offset reaction from vertical (degrees), as defined in Figure 10.1.

In the absence of laboratory data, a conservative value of 40° should be used for δ and 30° for ζ [100].

Scherle [97], in his comprehensive review on the subject, recommends values of face resistance factor (f_1) ranging from 300 to 600 kN/m² depending on soil type. The resistance force is found by multiplying the area of the excavation face by this factor, thus

$$P_S = \frac{\pi \cdot D_s}{4} \cdot f_1 \qquad \text{- eqn 10.3}$$

where the variables are described above. It must be judged whether the soil is likely to have a high or a low factor f_1.

The face resistance will vary considerably in different soils and, if the penetration mechanism is understood, its value can be estimated. The value will vary if different soils are encountered along the line of the drive, but will remain approximately constant for any one soil if the penetration mechanism remains constant.

Pressure balance operations. Here the face resistance will be equal to a combination of the contact stress applied by the soil to the cutting face of the machine and the hydraulic pressure of the fluid used to support the face. Ideally the contact stress between the machine and the soil should match the undisturbed soil stress, that is the lateral earth pressure at rest. If this is achieved, no ground movements towards or away from the excavation face will occur and thus no settlement or heave problems should arise as a result of excavation. The fluid pressure should be set to match the external water pressure in this case. Where there is no external water pressure, the sum of the fluid pressure and the contact stress should match the lateral earth pressure at rest in the soil.

The stress applied by the soil can be calculated with sufficient accuracy by simple lateral earth pressure formulae. The coefficient of lateral earth pressure at rest (K_O) for different types of soil is given in Table 10.2. For firm to hard clays (depending on their stability numbers), cemented soils, rocks and other self-supporting soils, the lateral stress acting at the face will be zero. However applying a contact stress according to the values in Table 10.2 will not be detrimental.

Table 10.2 Coefficient of earth pressure at rest for different soils

Soil type	K_O
Gravel	0.7
Dense sand	0.8
Loose sand	0.5
Stiff to hard clay	2.0-2.5
Firm clay	1.0-1.5
Soft clay	0.6-0.8
Silt, alluvium	1.0

A further method [98] is based on the N value as determined by the Standard Penetration Test (SPT). The shield resistance (in kN) is given by

$$P_S = 1.32 \cdot \pi \cdot D_S \cdot N \qquad \text{- eqn 10.4}$$

where the terms are as defined above. This equation yields the typical values shown in Table 10.3.

Table 10.3 Shield resistance (in kN) based on typical SPT N values

Machine diameter	SPT (N) value			
	5	15	30	50
0.60	12.4	37.3	74.6	298.6
0.90	18.7	56.0	112.0	448.0
1.20	24.9	74.6	149.3	597.2
1.50	31.1	93.3	186.6	746.4
2.40	49.8	149.3	298.6	1194.3

10.1.2 Line friction loads

The frictional force, known as the line friction load, acting on the pipeline as it is jacked behind the shield is influenced by many factors and is therefore difficult to calculate. It is perhaps for this reason that the UK pipejacking industry has typically used strict approximations, such as those reported by Craig [99] and reproduced in Table 10.4. These frictional resistances are multiplied by the external surface area of the pipeline to determine the jacking forces. The results of work [25] are reproduced in Table 10.4 also, and these data together show that very large variations in forces are recorded. It is clear that such imprecise empirical methods of analysis are of little practical value and that attention should be paid to the individual influencing parameters. The analytical equations used to predict line friction loads will be described in this section, assuming a straight drive with a uniform external pipe surface. The factors that influence the line friction loads, and should thus be considered in any predictive calculations, will then be described in section 10.2.

Table 10.4 Frictional resistance of pipes in different soils (after Craig, 1983, and Stein et al, 1989)

Soil type	Frictional resistance (kN/m²)			
	France	UK	Australia	Germany
Rock		2 to 3	1	
Boulder clay		5 to 18		2.8 to 18.4
Firm clay	8 to 10	5 to 20	5 to 7.5	5.3 to 9.3
Wet sand		10 to 15	13	2.2 to 16.1
Silt	17	5 to 20		4.9 to 8.5
Dry dense sand				1.1 to 6.7
Dry, loose sand	20 to 30	25 to 45		
Fill		up to 45		
Dense gravel	50			6.4 to 2.3

Stein and his colleagues [25] follow the approach of determining a simple frictional constant (M in kN/m²) in order to calculate the total resistance to movement of the pipe (P_p in kN) according to

$$P_p = M \, . \, \pi \, . \, D_p \, . \, L \qquad \text{- eqn 10.5}$$

where D_p = External diameter of the pipe (m)
L = Length of pipeline (m)

In its simplest form they recommend that a pipe-soil frictional coefficient, selected from a table or measured in the laboratory, is multiplied by the mean normal stress acting on the pipe wall calculated according to the principles expounded in section 8.2. The normal stresses acting immediately after construction should be carefully distinguished from those that develop in the long term. To complicate this calculation, various methods of calculating M are reported in which numerous external factors are considered. Care must be taken in calculating M, a method derived from the fundamental principles of physics being best.

Predicting forces in rock is apparently the easiest case of all since the excavation is necessarily greater in diameter than the pipeline, which thus rests on point contacts within the excavated cavity (Figure 10.1). The frictional resistance acting on the pipeline (P_p in kN/m run) is thus given by

$$P_p = \frac{W_p \tan \delta_p}{\cos \zeta} \qquad \text{- eqn 10.6}$$

where W_p = Weight of pipe (kN/m run)
δ_p = Coefficient of friction between pipe and rock (degrees)
ζ = Offset of reaction from vertical (degrees)

The forces will be small [100]. For drives below the water table in water-bearing rock, the pipeline will float and the forces, acting on the upper surface of the pipe, will be lower still.

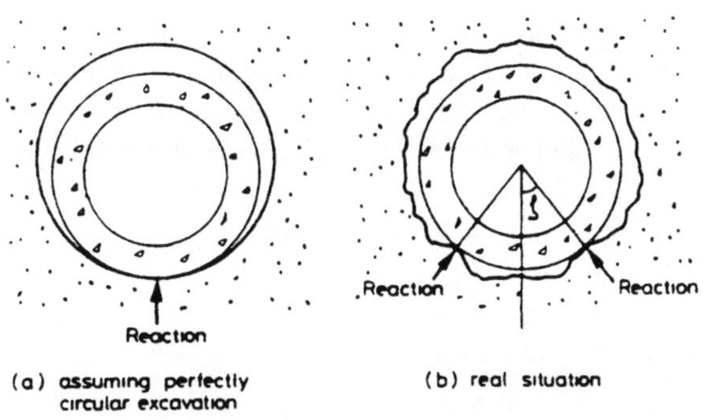

Figure 10.1 Pipe (or shield) seated in an oversized cylindrical void (after O'Reilly and Rogers)

In practice, however, the processes of excavation and, particularly jacking of the line can cause small rock fragments or rock flour to form. This material collects in the base of the drive and can force the pipe upwards, pinching it between the upper and lower surfaces of the void. As the rock fragments or flour are dilatant, this has the effect of filling the void and results in very high line friction loads as the pipe is forced into a progressively undersized void in strong materials. A second phenomenon noticed in practice is the mixture of rock flour with water to create a cementing paste. This similarly increases the forces.

In soils, it is far more difficult to predict jacking force, since these will depend not only on the type of soil and its variation along the drive length, but also on the time-dependent stability of the soil and the rate of pipeline advancement. The cover depth, position of the water table and any surface surcharge will also potentially influence the normal stresses applied to the pipeline, and hence the frictional resistance of the ground. Given this information and assuming that a standard overcut dimension of 10 mm is used in a perfectly aligned pipeline without lubrication, analyses can be attempted. Once these analyses have been made, the effects of varying the overcut ratio, misalignment of the line, use of lubrication and frequency and duration of stoppages can be determined by reference to work published on site experience.

A simplified approach has been suggested [101] in which earth pressure theory is used to calculate the forces acting normal to the pipe. By using free-field horizontal and vertical stresses, no allowance is made for overcut and the arching mechanisms that derive from them. The predicted forces would therefore be expected to overestimate those experienced in practice [102].

Herzog [96] adopts a similar approach in which the vertical stress (p_v) assumed to (p_h) is similarly the free-field stress determined using the coefficient of earth pressure at rest. The resistance to movement of the pipe (P in kN) is then determined using a frictional coefficient f_2, given in Table 10.5, in accordance with

$$P_p \quad = \quad \pi \, . \, D_p \, . \, L \, . \, f_2 \, \frac{P_v + P_h}{2} \qquad \text{- eqn 10.7}$$

where the parameters are defined above. In both of these simplified analyses, adjustments can be made for surface surcharge and the position of the water table. However, if the change in resistance caused by overcut is not taken into account, the analyses will produce overestimations of the resistance. Herzog's estimates are shown in table 10.5.

Table 10.5 Frictional coefficient (f_2) used by Herzog (1985)

Soil	Steel surface	Concrete surface
Gravel	0.55	0.88
Sand	0.45	0.65
Loam, marl	0.35	0.40
Low grade clay	0.30	0.35
Clay	0.20	0.25

Note: Low Grade Clay is assumed to mean a soft, alluvial clay

Jacking pipes have been modelled by considering them as an elastic cylinder resting in a cylindrical void in an elastic continuum [102] [103]. The model was applied to stable drives in London Clay, such that the pipes were resting in the bottom of the excavated void with a width of contact (b, in m) governed by elasticity. The *frictional* resistance (F in kN/m run) was then given by

$$F \quad = \quad \alpha \, . \, s_u \, . \, b \qquad \text{- eqn 10.8}$$

where α = Adhesion factor, commonly used in pile skin-friction formulae,

s_u = Undrained shear strength of soil (kN/m^2)

However, this equation was found to underestimate the forces when the pipe was considered in a case of full buoyancy, in which the pipe floated in bentonite lubrication. A range of forces based on different values of b, along with engineering judgement, should provide a resistance to adopt for predictive purposes. When the pipe is driven through a cohesionless soil, analyses based on a pipe seated in an oversized cylindrical void are inappropriate since the soil will collapse onto the pipe.

How fast the soil collapses will depend on any suction, or negative pore water pressure, in the cohesionless soil, although suctions will typically dissipate very quickly. The overcut on the shield will result in an active failure zone being created around the pipe, the extent of which will depend on the overcut ratio, the density of the sand and the dilatency of the granular material. Arching will redistribute the ambient stresses away from the pipe. The dividing line between cohesionless and cohesive soils is not, however, distinct and immediate behaviour in clays is very different to the longer-term case. Arching in heavily overconsolidated clays and stratified clays, and swelling of heavily overconsolidated clays, can considerably alter the assumed pattern of behaviour. These factors will be considered in more detail in the next section.

10.2 Factors influencing jacking forces

10.2.1 Soil stability and ground water

In addition to the basic differences in frictional coefficient of the soil itself, several factors influence the jacking forces in soil. The first, already mentioned, concerns the ability of the soil to support itself. Coarse non-cohesive soils will collapse onto the pipe either immediately or soon after excavation, since any negative pore water pressures created by the stress relief will rapidly dissipate. In fine-grained cohesionless soils, the permeability will be lower, and collapse will be delayed due to equilibration of pressures by water flow towards the area of negative pore water pressure. For clays, which are not only finer grained but also consist of complex clay minerals, the permeabilities are typically less than $10^{-8}ms^{-1}$ and stand-up times are considerable. This is why the pipe is assumed to be resting in an oversized void.

It has been commonly stated [104] [105] that full collapse of an unlined tunnel will occur at values of stability number (N_S) of approximately 6 and above, where in its simplified form

$$N_s = \frac{\gamma . h - \sigma_T}{C_u} \qquad \text{- eqn 10.9}$$

and γ = Bulk unit weight of soil (kN/m^3)
h = Cover depth (m)
σ_T = Tunnel support (internal) pressure (kN/m^2)
C_u = Undrained shear strength of soil (kN/m^2)

However the undrained shear strength applies to the measured strength of rapidly sheared clays and this strength will reduce with time, in effect causing the stability number to rise. In open face operations σ_T is zero at the face, whereas in pressure balance operations σ_T can often be adjusted to avoid stability problems at the face. Once the shield has passed, however, and the stability number is applied to the soil surrounding the excavated void, a similar situation occurs in both methods. In this case σ_T can only be applied via pressurised lubricant. If this is absent, time-dependent progressive collapse will occur.

In granular soils, arching takes place if a void is created because the material is invariably dilatant [106]. The degree of dilatency and the local stress regime will together determine whether significant overburden stresses are likely to be applied to the pipe. For example, the stresses applied to a pipe being jacked behind an oversized shield through a dense, angular, granular soil with a high angle of dilation will be far smaller than those in a loose, rounded aeolian sand, which will have a low angle of dilation. In the second case arching mechanisms are likely to break down altogether and full sand collapse might be expected.

Cohesive soils, however, can both contract and dilate, depending on the stress history of the clay. In soft, weak alluvial soils, collapse will occur easily since such soils lose volume, or contract, under shear and in effect flow onto the pipe. In heavily overconsolidated soils, which are dilatant, some degree of arching might be expected for the same reasons as for granular soils.

The arching mechanisms are induced as a result of the downward movement of the material immediately above the pipe, caused by the overcut. This material dilates, becomes looser and collapses. However, the material immediately overlying the pipe distributes the vertical stress to either side of the collapsing mass, commonly represented as a triangular block. Only a small overcut is required for this arching mechanism to occur. It could be argued that the stress relief at the face during excavation might be sufficient to induce arching. Recent research [107] has shown that significant draw along of (granular) material occurs due to friction at the pipe-soil interface. This tendency might well reduce the benefits of stress relief due to arching in cases where there is minimal (or no) overcut.

In granular soils, it would seem appropriate to specify an overcut of at least 10 mm (on diameter) to guard against force increase due to draw along [108]. In cohesive soils, only a small overcut is required; it is argued that the soil's inherent stability will generate relief while the stability number is suitably low. However in heavily overconsolidated clays, swelling will occur on stress relief as water is drawn into the clay by the high negative pore water pressures (or suctions). This can cause the stresses on the pipe to rise considerably unless significant overcut, typically 20 mm or more on diameter, is provided.

The position of the water table is another important factor affecting stability. The vertical stress in the ground below the water table is modified to

$$p_v = \gamma h_1 + (\gamma - \gamma_w)h_w + \gamma_w h_w \quad \text{- eqn 10.10}$$

where γ = Bulk unit weight of soil (kN/m³)
γ_w = Unit weight of water (kN/m³)
h_1 = Thickness of soil above the water table (m), and
h_w = Height of water table above point under consideration (m)

and this will influence the normal stresses acting on the pipe. The hydrostatic component, or pore water pressure ($\gamma_W h_W$), is omni-directional when assessing the stress field.

Equally important will be the facility to equalise pressure and the tendency of water to flow towards an excavation since the excavation will act as a (pressure) sink, or a drain, unless internal slurry pressures are used to maintain ground water pressure balance. The stability of non-cohesive soils will reduce since negative porewater pressures will not remain after excavation and seepage forces will act towards the pipe, tending to break down arching mechanisms. These factors will combine to increase the jacking forces. Similar, although less rapid, behaviour can be expected in clay. Such an effect, although for slightly different reasons, is reported in which jacking forces rose fourfold when dry conditions in a jack in stiff glacial clay became inundated [109]. In this case the combination of stress relief and available water was thought to result in swelling of the overconsolidated clay and closure of the overbreak.

More generally, the change in jacking force along the length of a drive will occur as a result of both change in face stability and change in frictional force. Assuming a constant rate of advancement, the face resistance will be soil material-dependent whereas the frictional resistance will be soil stability-dependent also.

An example of a jacking force record for installation of a 1.2 m external diameter, resin-coated pipe at a depth of approximately 6 m in alluvial clay is given in Figure 10.2 [110]. The most interesting point here is that over the first 150 m of the drive a silty clay crust with an undrained shear strength of 30 kN/m² was recorded in the borehole records above the pipe crown, whereas no such clay crust appeared in the borehole records for the second half of the drive. Thus in the latter half the soft to very soft clay, with an undrained shear strength of 10-20 kN/m², extended above the pipe crown, as well as occurring in the face over the full length of the drive. Stability numbers were approximately 3 and 6-12 for the first and second halves of the drive respectively.

It is apparent from Figure 10.2 that a constant face resistance of approximately 250 kN (25 tonnes) occurs by extrapolating the initial straight line back to the origin. Over the first 150 m of the drive the frictional forces, or line friction loads, cause a remarkably uniform increase of 1.1 kN per square metre of pipe surface. Between 160 and 290 m the force increases by 5.3 kN/m², which is almost five times as rapid an increase. This is attributed to the self-supporting nature of the alluvial clay in the early part of the drive and the collapse of the clay in the latter part. Collapse of the clay in the early part of the drive would have occurred at some time after excavation, and it is not clear whether this occurred while the latter part of the drive was taking place, although the remarkably uniform increases in force, at least up to 240 m, would suggest not.

Clay cap, a high density polymer flocculating agent that encapsulates clay particles, was used as a lubricant during the drive. However this does not appear to have masked the behaviour of the clay except that the measured frictional resistances might (and probably will) be considerably lower than they would be otherwise. To illustrate the simple frictional constant approach reported by Craig [99], the forces per surface area of pipe driven at various points along the drive have been marked in tonnes/m². These show a range of approximately 0.15 -0.30 T/m² (1.5 - 3.0 kN/m2) for the silty clay alluvium, whereas Table 10.3 indicates values of 5 - 20 kN/m² for both firm clay and silt. The use of a simple frictional constant, the range for which varies by a factor of 4 in any case, is shown to be misleading.

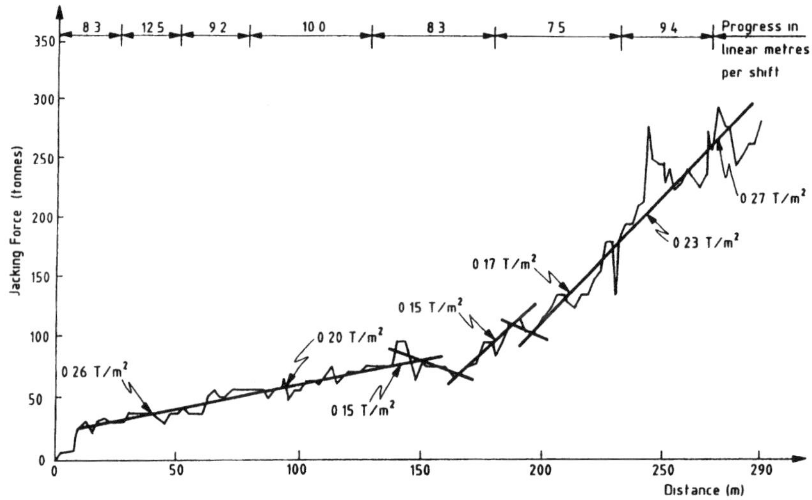

Figure 10.2 Graph of jacking force against distance driven for a pipejacking operation at Burnham-on-Sea (after Rogers, O'Reilly and Atkin, 1989)

10.2.2 Delays

After a stoppage, larger jacking forces are required to advance the line than are required to keep it moving. Although this phenomenon is widely reported, there are few published data to quantify this statement. This is partly because graphs of jacking forces are typically plotted for the cases where the forces have stabilised. For example, Figure 10.3 shows the forces at the end of each pipe driven for a 1.8 m diameter tunnel constructed using an Iseki Crunchingmole through alluvium [111].

It is clear from this figure that forces varied considerably even then, and that jacking forces commonly, though not invariably, rose after weekend stoppages. The jacking records show that restart forces were typically 50% higher, and sometimes more. For example after the weekend of 11-12 May a restart force of 860 tonnes was required to start the pipe moving, this had reduced to 340 tonnes by the time that one new pipe had been driven. It is therefore recommended that a capacity of twice the baseline force be available to advance a pipe train following a substantial stoppage. This capacity will be required of both the main rams and the thrust wall. Alternatively it may be necessary to use IJSs to cope with high start up loads.

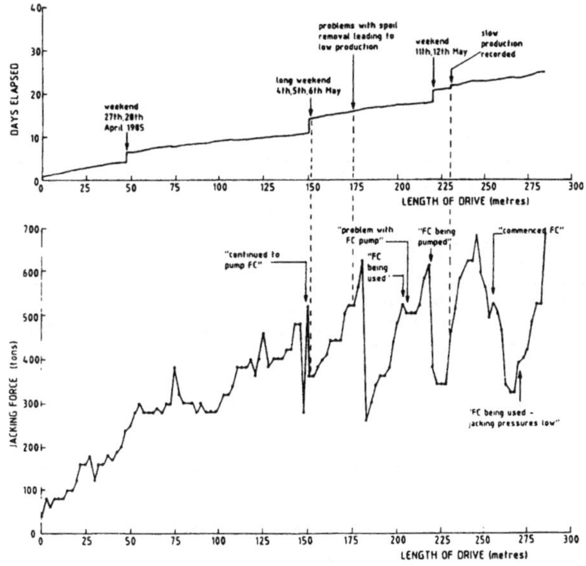

Note: FC refers to friction cut, a lubricant

Figure 10.3 Graph of jacking forces against distance driven for pipes jacked behind a 1.8 m diameter Crunchingmole through alluvium at a depth of 10 m (after Rogers et al. 1989)

A more detailed study using instrumented pipes in clay [78] has shown that the effects can be considerable even for relatively short stoppages, and that the effects are most noticeable in highly plastic clay (Figure 10.4). In this case the increase in force required is both rapid and repeatable, and can be attributed to the fact that high pore water pressures are generated by the undrained shearing associated with advancing the line. When jacking stops, the shear plane consolidates under the (in this case high) normal total stresses acting in the soil adjacent to the pipe. As a result, the high positive pore water pressures dissipate and the effective radial stresses, which control the frictional forces experienced by the pipe, rise. When jacking starts again, frictional forces are greater until the pore water pressures caused by undrained (or partially-drained) shearing reduce the effective normal stresses once more. It is interesting to note that jacking jobs have been successfully restarted several months after stopping work. Having broken the initial bond, the jacking loads seem to revert to a normal level, much as they do after shorter stoppages.

10.2.3 Lubrication

Lubrication and lubrication systems can take many forms according to the job and the nature of the soil conditions. Theoretically, if an appropriate lubricant is introduced and can be maintained in the annular space between a pipe and a stable soil wall of the bore, the pipe being jacked can be suspended in this fluid. The force required to jack the pipe will then only be that required to overcome the shear value of the fluid. It has been calculated that in such a condition it would be possible to push a length up to 1000 times greater than without the fluid for the same load.

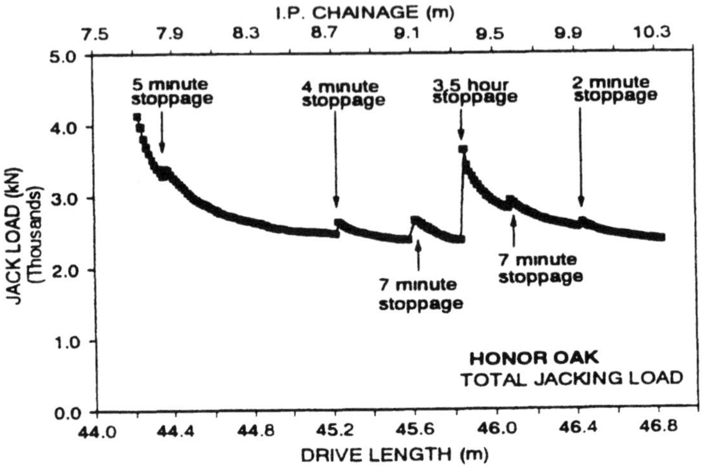

Figure 10.4 Detailed variation of jacking forces with distance driven for a deep hand-excavated drive in London clay (after Norris and Milligan).

In practice, in cohesive soils (e.g. stiff clays) the ideal situation can be more closely realised. With expansive clays, in stress-relieved strata and generally in weaker clays and silts, it will be hard to maintain the initial bore because it is difficult to place and maintain a suitably pressurised fluid in the annular space. The inward movement of material onto the pipe will result in increased contact and frictional loads. Once the soil has moved inward, to force it back and relieve the resistance by injecting grouting becomes a difficult, although not impossible, task.

In non-cohesive soils the problems of maintaining the annular void are much greater, as the slurry will be lost into the permeable soils. In such soils the aim is to build up a filter cake which will minimise the fluid loss. The layers of soil adjacent to the pipe will become a soil/lubricant mixture which will help reduce frictional forces.

Lubricating fluids. The fluids used for lubrication must have certain inherent qualities. They must be non-hazardous, environmentally compatible, and stable for their design life. Lubrication fluids have also to fulfil a number of functions over and above that of simply reducing friction between pipe and ground These can include:

- Filling the annular space between the diameter bored.
- Filling overbreak or voids outside the bore.
- Ensuring that soil does not adhere to the pipe or casing.
- Providing a stabilising medium to prevent the collapse of soil onto the pipe.
- Providing a stabilising filter cake to the soil face of the excavated bore.
- Reducing water inflow.

The main carrier fluids are water and diesel, but it is the water-based fluids that are used in pipejacking and microtunnelling: water, bentonite, bentonite-based formulations and polymers Water is frequently used as a lubricant, particularly for working in clays and silty clays. The injection of water at the head can be beneficial. However, the injection of water round a bore can lead to settlement or other geotechnical problems in the surrounding soil caused by softening of silts and clays, squeezing, or in the case of some absorbent clays, expansion.

Bentonite is the popular name given to the natural clay materials, potassium, calcium and sodium montmorrillonites, all widely used in the petroleum, tunnelling and civil engineering industries. Their prime values are their capacity to absorb water (basis of yield) and their thixotropic characteristics, the property of becoming temporarily liquid when shaken or stirred but returning to a gel consistency when at rest. The sodium montmorrillonites found in Wyoming, can absorb up to·14 times their original volume. Many industrial bentonites commercially available are of comparatively low grade and will have a low yield compared to that of better materials such as the sodium bentonites. To maximise yield, the water should be pre-treated for hardness and pH. High grade bentonite will provide a lubricating fluid of the required viscosity, a degree of water loss control, a supporting gel and a filter cake which will help stabilise the bore wall and unconsolidated formations. The bentonite provides a reasonable level of lubrication. Being a natural clay material it is environmentally compatible. However, these qualities of a high grade bentonite, with the possible exception of its environmental compatibility, can be enhanced by the addition of other chemicals.

Bentonite-based mixes aim to improve water-loss control, to increase viscosity and provide lubrication. Water-loss control reduces the chance of clays absorbing surplus water, swelling and becoming unstable, all of which will increase friction on the pipe. Water-loss can be reduced by adding various polymers, typically a polyacryllamide. Other types of additive available are lignins and organic polymers. Viscosity can be provided by the bentonite alone, but the quantities required are uneconomic. If higher levels of viscosity are required, small amounts can be added of chemicals such as the polyacryllamides, lignins and organic polymers, which will increase the viscosity of the bentonite. Polymer bentonite mixes of this type can stabilise clay formations and bind the surface of unconsolidated formations. The addition of some long chain polymers can improve the quality of lubrication. Further improvement can be made by adding specific lubricants which will mix in water-based fluids. Some are oil- or fat-based and act as a surface-active agent. Others are mechanical in nature and act like tiny ball bearings to provide the lubricity.

Polymer-based lubricants provide viscosity and filtration control as well as lubricity. Typically these are anionic polymers, synthetically produced. In practice, these have proved difficult to control in site conditions and are therefore usually used in low concentrations with high-grade bentonite.

Lubricating equipment and methods. Lubricating mixes are prepared in a colloidal or other form of grout mixer. It is good practice to allow bentonite mixes to be held in a storage tank for some hours before use. With a suitable grout pump, the fluid can then pumped into the line through a temporary pipe system connected to sockets cast into the pipe. The spacing and number of sockets should be designed to ensure that the lubricant is distributed evenly over the whole drive. Figure 10.5 shows a typical arrangement for lubricating pipework. The injection pressure will have to be varied according to lubricant, length of drive and soil conditions, but generally a lower pressure will allow a steady flow of material and a more uniform distribution. Only on longer drives should pressures over 50 bar be needed. It is important that provision be made for monitoring pressure both at the pump and at the injection points. The volume of material injected should bear a relationship to the overcut volume and also an estimate of the loss into surrounding soil. The volume of injected material must also be monitored and measured.

In microtunnelling it will not be possible to install and remove a temporary piped injection system in pipes less than DN 750 mm. However, some lubricant can be injected around the rear of the shield. An exception to this is with two-pass systems, where the temporary steel casings provide a relatively easy means of providing the pipes and sockets for a line lubrication system.

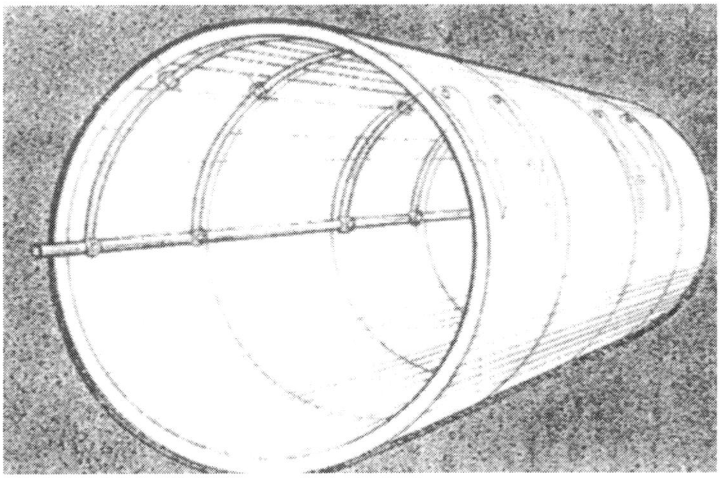

Figure 10.5 Layout for lubricating pipework.

Effects of lubricants on load reduction. Considerable variation in effectiveness has been reported. For example, Norris and Milligan [109] report a variation in apparent coefficient of skin friction by a factor of 3, ranging from full effectiveness to no effect, depending upon the composition of a silty sand and the injection procedures used.

A reduction in jacking forces of more than 50% is commonly quoted when lubrication is applied to a jacked pipeline, and an example is given in Figure 10.3. Takeshita [112] describes tests on jacked box sections which indicate a consistent reduction of 20% in jacking forces where bentonite was injected around the box section, and 40% where bentonite lubrication was used in conjunction with an overcut. Ishibashi [113] reports reductions in forces of 30-50% in clay and 20% in sands. However these figures are both site and process specific and, although encouraging, are of qualitative value only.

As already noted, the role of lubrication in cohesive and granular materials is likely to differ considerably. In a strong, self-supporting clay, lubrication serves two immediate purposes as well as potentially supporting the clay in the longer-term as a result of fluid presence and/or pressure in the overbreak.

It has been suggested [102] that the pipe will become buoyant if the annular space is filled with lubricant, and thus the frictional force will reduce. In this way the pipe self-weight, acting as a normal stress over the contact area at the *base* of the excavated void, will be replaced by a lower buoyancy stress acting over the contact area on the *upper* surface of the void. The second effect is that of a shear zone occurring within the low-friction lubricant zone or within the lubricant-clay mix at their interface.

In a granular soil, the lubricant mixes with the soil to produce a stabilised gel that prevents total collapse and provides a low-friction material through which the shear zone forms. Fluid pressures are also expected to form, their dissipation being restricted by the thixotropy of the fluid. This gel layer has been shown to be approximately 10 mm thick and to produce buoyancy when properly injected all round the pipe [114]. On a microtunnelling project in sandy-clayey soil, the jacking forces remained low and approximately constant throughout the drive [115]. This was attributed to gel formation. Data from the properly lubricated pipe-jack referred to above confirm this view; the forces remained approximately constant in loose sand and gravel below the water table [114]. Similar observations in a drive through glacial clay with lubrication are shown in Figure 10.6. In the first half of the drive, through alluvium, the forces rise approximately linearly. Once the glacial clay is reached, the forces remain approximately constant, possibly indicating that the pipe is floating and that a low friction interface has been created.

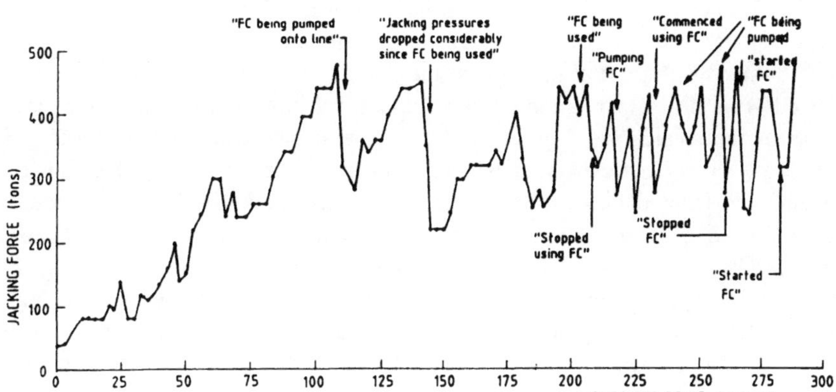

Note: FC refers to friction cut, a lubricant

Figure 10.6 Graph of jacking forces against distance driven for pipes jacked behind a 1.8 m diameter Crunchingmole through alluvium (to 150 m) and thereafter glacial clay at a depth of 10 m (after Rogers et al.)

10.2.4 Coatings and membranes

Coatings. The type of material used to construct the external surface of the pipe influences the jacking forces by its direct frictional resistance. This interface can be tested in the shear box to determine frictional coefficients, which must subsequently be modified to account for site conditions and construction operations. A smooth external surface is important; lower forces have been observed in cases where a concrete pipe has been coated with, for example, a resin-based corrosion protection system. A strong material is less likely to have its surface finish eroded or scoured during jacking, and is therefore likely to generate lower frictional forces. An additional factor is that of porosity and the material's ability to provide a drainage path to the surrounding soil. This is most important in clays, where drainage of excess pore water pressures can increase jacking forces markedly even after short stoppages.

Slip coatings can be applied to the wall of pipes as they pass through the drive pit. It is claimed that they significantly reduce friction and thus jacking forces. Materials used include clay cap polymer. This is a synthetic, anionic, free-flowing powdered polymer, which can be mixed with fresh or sea water in varying proportions, but typically 1 to 10.

Membranes and drag sheets. An alternative means of reducing friction is to introduce an artificial skin to separate the pipe being pushed from the soil. This idea was first introduced more than 20 years ago in the UK in the form of *drag sheets*. By feeding a flexible membrane (rubber or synthetic belting or thin steel sheeting) through the back of the shield and over the top of the pipe, it is possible to provide an ideal interface between soil and pipe. The membrane remains in contact with the soil and a lubricant is introduced between the pipe and the membrane. This reduces frictional load considerably.

This concept has since been extended in a Japanese technique called *membrane wrapping*. Rather oddly, the stated main purpose was not to reduce jacking loads but to provide a waterproof external lining to the pipeline in areas with high ground water. A complete sheath is fed out of the back of the shield to totally encompass the pipe [116]. A laminated vinyl cloth sheet is stored in a compressed form in a container pipe attached to the rear of the shield (see Figure 10.7). As the shield advances, the sheet is fed out, enclosing the pipe. Lubricant is introduced to maintain the annular gap and thus to protect the membrane from damage by the pipe as it is jacked forward. A high concentration lubricant is used (110 kg. bentonite, 0.75 kg high water absorption polymer to 1 m^3 water).

Toyhama illustrated the reduction in jacking load by comparing the jacking forces for two conventional jacking drives to three sections driven using membrane wrapping. In each case a 1000 mm concrete pipe was jacked using a slurry shield with gravel crusher. The work was below the water table in a sandy gravel (58% gravel, 31% sand and 11% silt) with a maximum particle size of 250 mm. Figure 10.8 shows the results and the overall reduction in jacking forces. For example the Nagai No 3 drive recorded only 34% of the conventional maximum thrust force (maximum jacking load 143 tf against 420 tf). Interestingly, the increase in jacking force at the beginning of all the drives was generally the same for conventional and membrane-wrapped. Loads for the wrapped pipes did not increase with drive length. After reaching a peak at around 20 to 30 m into the drive, the loads plateaued or even fell slightly. Trials in sandier soils indicated that a wrapped drive generated about 75% of the normal load required by conventional pipejacking.

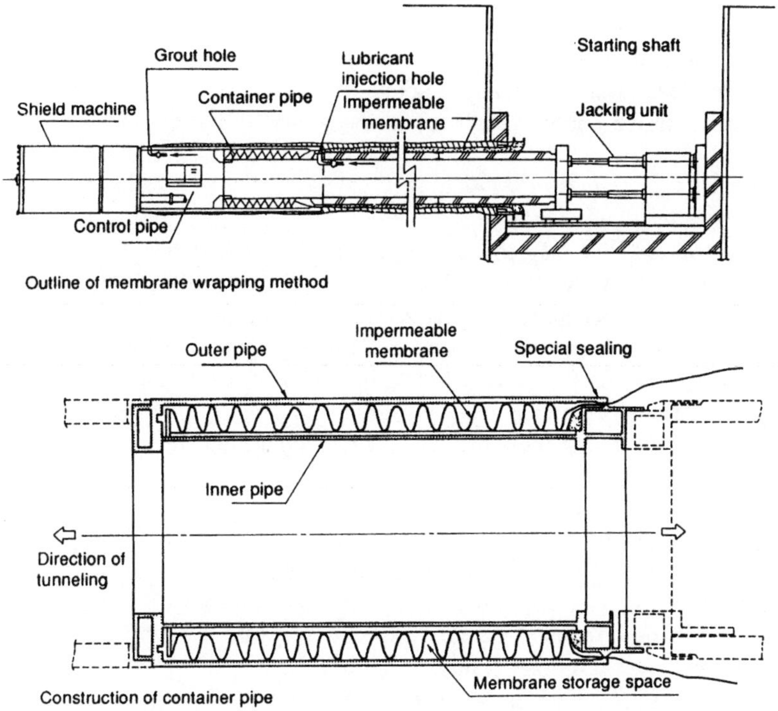

Figure 10.7 Membrane wrapping shield (diagram after Toyhama).

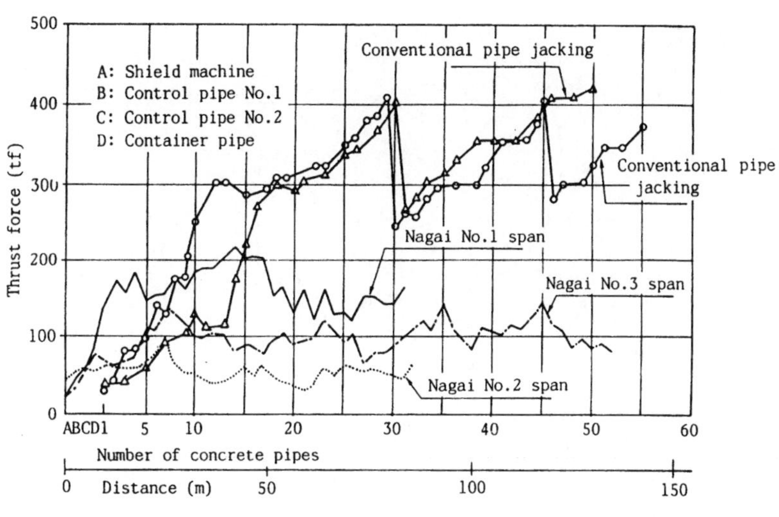

Figure 10.8 Nagai drives with and without membrane wrapping (after Toyhama)

Intermediate jacking stations (IJSs). The use of IJSs is often referred to in the context of reducing the jacking load, but this is a misunderstanding. IJSs reduce the maximum stress by distributing the load over a number of points in the line, but the jacking load is not reduced. It remains the sum of the individual jacking points.

10.2.5 Overcut ratio

The influence of overcut ratio, as defined in Figure 10.9, on soil stability has already been discussed in some detail in section 10.2.1, so only its direct effect on jacking force magnitude will be described here. Typical values of 10-12 mm overcut on radius are commonly quoted, although overcuts as large as 75 to 150 mm on radius have been used. Where large overcuts are used, the void has to be backfilled by grouting once driving is complete. This additional operation can only be justified if (a) considerable savings in production are achieved during the work and (b) no significant settlements occur during construction.

In a research project at Loughborough University of Technology, various overcut ratios between 0 and 0.140 were used in tests with 200 mm diameter pipes in both a uniform and a well-graded sand [118]. This work indicated that the optimum overcut ratio was 0.04. The forces remained low as the overcut ratio was increased above this value, but rose sharply as the overcut ratio reduced to zero. This optimum value equates to 20 mm on the radius of 1.0 m pipe, and 40 mm on radius for a 2.0 m pipe. It should be noted that these values are optima for sand above the water table and relate to the generation of arching. There are no data to suggest that an overcut of this magnitude is either an optimum or of value in cohesive soils, except perhaps in heavily overconsolidated clays that are liable to swell during the jacking process.

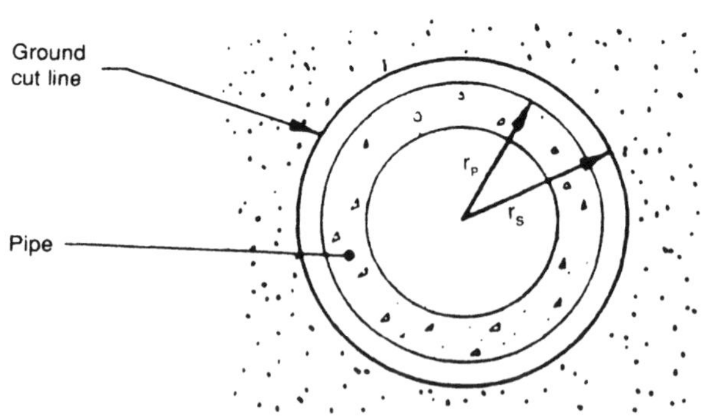

$$R = \frac{r_s - r_p}{r_p} = \frac{D_s - D_p}{D_p}$$

Figure 10.9 Definition of overcut ratio.

Takeshita [112] examined the forces required to jack rectangular box sections into sands and silts of different density, both with and without steel plates of different thicknesses welded to the upper and lower surface to provide an overcut. He concluded that a 50% reduction in jacking forces can be achieved using an overcut and that the optimum overcut is 1.2% (equating to an overcut ratio of 0.012) by consideration of equivalent diameters.

Above this value the forces remained the same at an overcut of 0.024, but rose slightly for an overcut of 0.036. It should be noted, however, that these box sections were combined to create a structure by jacking new sections immediately adjacent, and that the arching mechanisms will be different to those of circular pipes.

10.2.6 Pipe external surface

The external surface of the pipeline is important in terms of the material that forms the outside diameter of the pipe and also the joint alignment. The joint profile and the possibility of a step at the joint provide a potential bearing area against which the ground can act directly, as shown in Figure 10.10. Steps of more than 25 mm are possible in DN 1500 mm pipes manufactured to some official pipe standards, for example, ASTM C76-90. Any misalignment can result in potentially large force increases, since the load transfer mechanism is one of direct bearing rather than friction. Three cases should be considered:

- Joint profile when pipes are perfectly aligned.
- Joint profile when the pipe train is misaligned, but there is no lateral displacement of the pipes.
- Joint profile where lateral displacement of the pipes occurs.

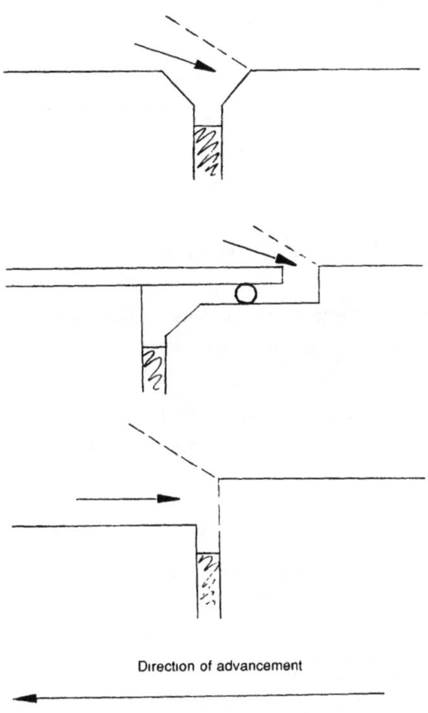

Figure 10.10 Effects of joint detail and joint misalignment on jacking forces.

Tests have been conducted on concrete pipes with an open joint detail and steel pipes with closely fitting joints created by careful machining [118]. The difference in interface friction between concrete and steel surfaces was found to be surprisingly small in shear box tests; the steel had a slightly lower value. However, the difference in jacking forces in the model pipe tests was considerable. The forces required to jack the concrete pipes were 2.1 - 2.7 times as great as those required for the steel pipes, and this can be attributed mostly to the difference in joints. The degree of misalignment in these tests was small and, although possibly significant, it would have influenced both pipe trains causing a slight increase in joint gap in both cases. In these tests there were no steps due to lateral displacement at the joints, but it has been suggested that steps will increase the jacking force considerably.

The increase in forces at joints is difficult to approach theoretically. The resistance at aligned joints (with no step), could be calculated using soil parameters alone in the joint vicinity, rather than interface frictional parameters, once the line is moving and shear planes have been formed. The effect of the open joint in this case would be to displace the shear plane away from the pipe wall at joints in dilatant, especially granular, materials. When jacking starts the exposed joint will act as a bearing area initially, or in effect an area of roughness to which the soil adheres, and thus shearing of the soil will necessarily occur.

Where there is a stepped joint, the bearing area is directly, rather than obliquely, opposed to the direction of jacking. In attempting to predict the increase in forces, it could be assumed that there is an effective increase in face resistance, the resistance being given by

$$\text{Increase in force} = \frac{A_{step} \cdot P_s}{D_s \; t_s} \qquad \text{-eqn 10.11}$$

where A_{step} = Area of step at the joint (m^2)

P_S = Face resistance determined, or measured, when shield passed that point (kN),

D_S = Diameter of shield (m), and

t_S = Thickness of shield (m)

This is likely to be an underestimation of the force because the soil is inherently confined around the joint, whereas at the face there is stress relief and a lower force could be expected. The situation here is more similar to that of the end resistance to a driven pile. Excessive loading at a step would result in spalling but no effective reduction in bearing area. Once the line is moving, the shear plane will form a smoothed discontinuity.

10.2.7 Misalignment

An increase in the jacking forces results from increased radial stresses on misaligned pipes. This can be masked by the scatter in jacking forces that arise from other influences as well. Haslem was the first to address the problem of alignment in 1986. He considered the pipeline to be a continuous structural unit and thus the radial stress to occur on the external surface of a curve. Later work on jacked pipes that were maintained in deflected positions demonstrated that large loads were induced in the lateral support systems used to maintain the deflection [119] [120]. These observations related to two adjacent pipes, loaded as though they were in a pipe train. It was demonstrated that discrete rigid elements will attempt to realign when the axial force is applied, which will result in large radial forces on the *inside* surface of a curve. This later observation, opposed to Haslem's, resulted from considering discrete rigid elements rather than a continuous pipeline.

In instrumented field trials on misalignment [114] good agreement was found between the maximum angular deviation of the pipe train and the positions of peak radial and shear stress. An example of radial total stress plotted against angular deviation is given in Figure 10.11. Although revealing much scatter, this indicates a significant trend. The results of these trials confirmed that the pipeline tries to straighten at positions of maximum misalignment and thus radial stresses occur on the inside surface of a curve, the pipeline behaving as a post-tensioned tube. The average radial stresses acting on a pipe at the point of maximum misalignment were found to be several times greater than the average radial stresses for the pipeline as a whole. Because of the position of the instrumentation used in this work the maximum radial stresses were probably not recorded; local concentrations were likely to have been greater.

Predicting the forces caused by misalignment is difficult due to the various factors described above. Published jacking records suggest that vertical misalignments result in greater force increases than horizontal misalignments. If the major effect is the increase of radial force, the most logical adjustment is to assume an effective increase in pipe length at an average pipe radial stress. From the limited data available, a threefold increase in radial stress for every 0.1 degree of relative angular deviation appears appropriate in dense or heavily overconsolidated soils. A significantly lower value would apply in less competent soils. Thus for a pipe at an angular deviation of 0.2 degrees in a dense soil, the force required to advance the pipe would be approximately six times that for properly aligned pipes. Further work on this subject is needed before confident predictions can be made.

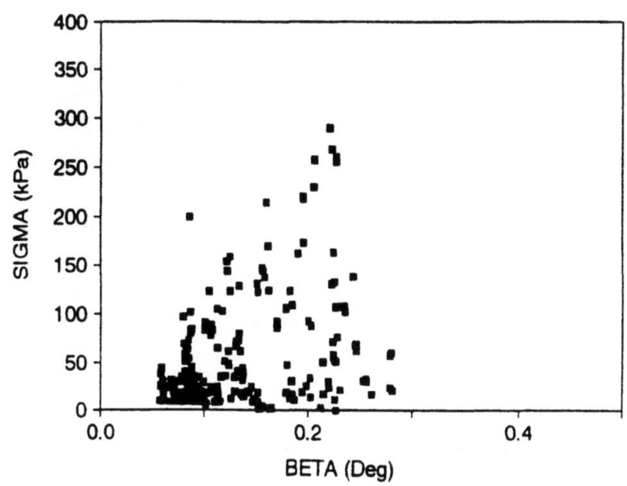

Figure 10.11 Relationship between angular misalignment of a pipe train and total radial stress (after Norris and Milligan)

10.3 Pipe loads

The forces required to advance the pipeline are applied to the pipes by the main rams in the jacking pit and, where necessary, by additional rams housed within IJSs. These forces act through a steel ring that spreads the point loads from the rams over the full cross-sectional area of the pipes. If the pipes were perfectly uniform in both material and shape, were perfectly aligned longitudinally and laterally and were in contact with both each other and the load spreading rings via perfectly smooth faces, then the load transfer through the pipes would be uniform around the cross-section and would reduce with distance from the jacking pit due to friction.

At relatively small angular misalignments (for example, 0.025°), the load transfer at joints has been shown to be controlled by the material properties of the packing, but is almost totally dominated by the position and magnitude of the angular misalignment when larger.

Research suggests that the choice of packing material is highly important in reducing the tendency to stress concentrations in drives at typical angular misalignments, say of 0.05° or less. Suitable packing might help limit the maximum stress concentrations at higher angular misalignments. Packing materials should have certain characteristics. First, so that they generate little or no radial stress which could cause spalling at the edge of the joint, they should have a Poisson's Ratio under compression of approximately zero. Second, they should have the ability to compress greatly, although not completely, when loaded but to recover their thickness when unloaded. For this reason rubber (although perhaps better able to provide an appropriate axial stress distribution) would prove unsuitable, since it has a Poisson's ratio of approximately 0.5 and would generate very high radial stresses. In practice various types of wood or fibreboard packer are used. As this material has shown considerable permanent deformation when the first load is applied, the design should account for its precompressed stress-strain characteristics.

Several types of timber have been shown to be inappropriate, and thicker packers to perform better than thinner ones [122]. The compression characteristics of the three favoured materials set out in Figure 10.12 indicate that the dense fibreboard compressed most and consequently performed best. The recovery of the materials was greater (although maximum compression remained unaltered) when they had been saturated and allowed to dry. Compressibility was also greatly enhanced when the materials were saturated. The researchers concluded that the packers would distribute stresses up to an angular deviation of 0.2° under ideal conditions.

A later study [123] produced similar findings, with nylon packers proving too rigid and timber packers proving suitable. Both the relative rigidity and relative hardness of the packers in relation to the pipe were stated to be important. To use new materials with a negative Poisson's Ratio was put forward as a possibility.

10.4 Curved drives

The advent of curved drives for pipejacking and microtunnelling has removed one of the limitations of the technique. When designing a curved drive, the following factors need to be considered:

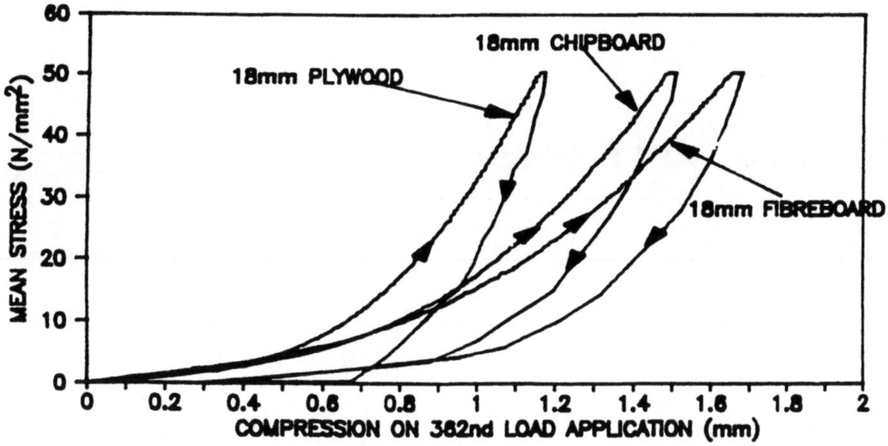

Figure 10.12 Relationship between stress and compression for packing materials (after Milligan and Ripley)

- Radius of curvature
- Individual pipe length
- Means of transfer of jacking forces
- Pipe diameter
- Maximum and minimum joint gap
- Maintenance of seal between pipes.

Experience suggests that joint details and pipe length are critical for curved jacking, and that short lengths are likely to be preferred. Offset against this requirement will be the increased production costs. Pipes with skew end planes at one or both ends permit tighter curves (100 m and less), while providing better joint load transfer and watertightness [124].

Atkin [125] describes curved jacking with radii down to 300 m using short (1.18 m), square ended lengths of 1.5 m diameter pipe through water-bearing sands and gravels. The joint opening for these pipes was reduced from 15 mm to 7 mm which, with 19 mm thick plywood packers, ensured adequate load transfer. A plot of jacking forces against distance driven indicates that large forces were not required as a result of the curvature.

The membrane wrapping method described by Toyhama [117] has been used for driving curved installations with 1100 mm concrete pipes at a maximum radius of 200 m. Driving tighter radii than 200 times diameter will require special provisions. The Unitunnel system described in Chapter 4, with a jacking station at each joint, is used in Japan for installing concrete pipes in the range DN 800 to 2200 mm. Curves as tight as 31 m for DN 1100 pipes and 40 m for DN 1500 mm pipes are reported. Showa Civil Engineering, a Japanese contractor, uses a method called the LDCI System. This system involves injecting a high density mud slurry into an overcut around the tail of the shield, allowing the following pipes to move through the curves. Curve radii of less than 100 m have been driven for installing pipes in the range DN 600-900 mm.

11 Drive and reception pits

11.1 Dimensions

11.1. 1 Drive shafts

The dimensions of shafts relate to the type of equipment being used. The width is in direct relationship to the diameter of the pipe being driven, plus an allowance for working space. External diameter of pipe plus 2000 mm is typical. The length is determined by the length of pipe sections to be installed and the space needed for the jacking equipment. Attention also needs to be paid to the longest shield section which has to be handled in the pit, as this can be greater than the pipe lengths. The most compact microtunnelling equipment will work in a shaft the length of the pipe section plus 1000 mm. Some Japanese equipment may require anything up to 5000 mm. For pipejacking in diameters up to DN 1500, the minimum additional length to be added to the pipe length increases from 1500 mm to 2000 mm. Shaft dimensions for drives in larger diameters tend to be site specific. Manufacturers provide information on dimensions of shafts required for their equipment.

11.1.2 Reception shafts

The dimensions of reception shafts must provide adequate working space and also allow for the recovery of shield components, or temporary sections in the case of a two-pass installation. Some manufacturers now make their shields in modular form so that they can be recovered from smaller shafts.

11.2 Shaft construction

Choice of shaft construction for driving and reception pits will depend on the depth required, the size of equipment and pipe sections, soil conditions, ground water, local custom and safety regulations.

11.2.1 Stable ground and shallow depths

In stable soil conditions, drive shafts down to depths of 6 or 7 m can be constructed with various techniques. Excavation, sheeting of the sides and strutting are undertaken at the same time. Safety regulations require that excavations and working shafts of this type are properly sheeted and framed. In the USA, the design of shafts for most jacking work has to be undertaken by a registered designer [126].

Design. Thrust shafts and reception pits are designed in the same way as any other excavation, accounting for superimposed loads and traffic loads as well as those from the soil and water pressure. In addition, allowance must be made for the jacking loads that are transmitted to the rear wall and pit strutting.

Standard sheeting. Where the sheeting is taken down as the excavation proceeds, timber and trench sheeting are widely used. Figure 11.1 shows a well-constructed two-stage shaft using sheeting, whalings and struts. As sheeting in these materials does not provide much of a barrier to water or seepage of fine particles, they are more suitable in good cohesive soils. Fully-interlocking sheet piles offer greater strength and a better barrier when one is required.

Figure 11.1 Jacking shaft constructed with two-stage sheeting, whalings and struts (photograph by courtesy of Euro-Iseki Ltd.).

System sheeting. Prefabricated shoring boxes with hydraulic struts are now common for shaft construction. The package includes pit corner posts with slide rails into which the sheeting is guided. A basic depth of construction of around 3 m is possible with a single-stage box. If a deeper pit is required a second sheeting rail is provided which allows a second set of sheeters to be added.

Liner plates. Liner plates are made of steel, often galvanised, and come in many sizes, radii and curves. These can form circular or rectangular shafts with straight sides and curved ends, as shown in Figure 11.2. In stable conditions, the excavation is taken down a sufficient distance to allow the next plate to be placed and bolted to the plate above. So that it can assist with temporary strutting, the central core of excavation is left in until the ring is complete.

Figure 11.2 Standard size shaft constructed from liner plates (photograph by courtesy of Sanwa Kisai).

The Japanese favour rectangular pits with rounded ends made to standard sizes with corrugated liner plates for microtunnelling and jacking equipment, whereas in North America flanged liner plate is popular for sinking circular shafts. Again the procedure with stable ground is to build from the bottom down with sufficient excavation to allow each new section to be bolted in. This is known as *underpinning* or *chimney* construction. The same type of liner plate is used in caisson sinking.

In Europe precast concrete segments are used as standard liners. There are two alternative systems: bolted and smooth bore. With these types of segment, shafts can be constructed by the underpinning or caisson methods. Traditional segments can be bolted together through the internal flange on all four edges and a watertight seal can be incorporated at the joint faces. A bolted segment shaft for a pipejacking installation is shown in Figure 11. 3. Shaft diameters may range from around 1500 mm to over 10 m

Figure 11.3 Shaft formed from bolted segments (photograph by courtesy of Euro-Iseki Ltd.).

One-pass segments have the advantage of providing a smooth interior wall to the shaft. Vertical holes are cast centrally into the segment and tie rods in these holes connect one segment to another on the horizontal faces. Special T bolts lock into a socket cast in the ring above. Connecting adjacent segments at the vertical face is at a small pocket. Each segment has alternate single and double steel loops which project into the pockets so that when segments are brought together, the single loop on one segment interlocks with the double loop on the other. A central core fastener is driven home to stress the loops and make the connection. Figures 11.4 shows the forms of fixing

11.2.2 Unstable ground and deep shafts

The theory and practice of deep shaft construction in difficult ground is an extensive and complex subject. There will be many cases where specialist advice will be required for the design, and expertise for the implementation [92]. Complications aside, there are three basic approaches: cofferdams, caissons and advance stabilisation.

Cofferdams. A cofferdam is a temporary structure designed to support the ground and to exclude water (or as much as possible) from an excavation. Essentially a contiguous wall is constructed in the soil in advance of excavation. The best known of the many forms of cofferdam is the sheet-steel piled cofferdam, where interlocking sheet-steel piling is driven down well below the base of the shaft. This construction, commonly used for both drive and reception pits, has two main advantages for pipejacking. Firstly, pits can be made rectangular in plan to allow the use of long-length jacking pipes. Secondly, in very weak ground, sheet piled pits can be made wide enough to generate the jacking forces required for large diameter pipes or long pipe strings.

The design of sheet-steel piled cofferdams is well established. M.J. Tomlinson's *Foundation design and construction* [92] is a useful guide to its theory and practice. A typical sheet-piled pit is shown in Figure 11.5.

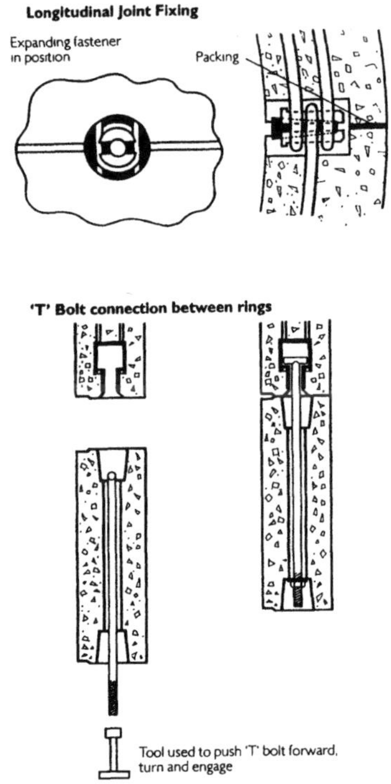

Figure 11.4 Connection details for one-pass segments (diagram by courtesy of Charcon Tunnels Ltd.)

Although sheet pile cofferdams will continue to be used for long-length pipes and for large diameter pipejacking, they are not economic for most microtunnelling and smaller diameter pipejacking jobs. Steel-sheet piles are expensive and to this outlay must be added the costs of transportation and the mobilisation of pile-driving and extraction equipment. Apart from cost, another disadvantage is that it is often difficult to sink a sheet-steel piled shaft in an urban location without first diverting existing utility lines. The shaft must also be backfilled and the sheet piles and framings withdrawn once the drive is completed. The vibration and the voids that occur during extraction can cause ground and property settlement.

When working below the water table, there is a risk of failure due to hydraulic uplift (boil or heave) during the excavation stage, which can lift the concrete floor of the shaft after it is placed. To avoid this uplift, the penetration of the piles below the base level is critical. Their cantilever effect provides resistance to the tendency towards inward movement of the piles while the cofferdam is being excavated. In most soil types any vertical seepage can be controlled by *sump-and-pump* techniques. However, in more permeable soil it may be necessary to employ well-point dewatering or pregrouting to control seepage.

Alternative methods of forming the cofferdam include secant piling and diaphragm slurry walling, but are only occasionally used for jacking.

Figure 11.5 Sheet steel piled jacking shaft (photograph by courtesy of Laserbore Ltd.).

Caissons. The essential feature of a caisson is that the circular structure is constructed above ground and then sunk as a single unit to the required depth, where it remains part of the permanent work. Excavation, progressively carried out as the caisson is sunk, is often under water and working may sometimes have to be in compressed air.

Caisson techniques, used for nearly 150 years in the construction of deep foundations and shafts for tunnelling and mining work, have been used on a smaller scale for both pipejacking and microtunnelling. Caisson sections for microtunnelling are more likely to be precast and delivered to site. For larger installations, sections may be cast on site, as were 5 m diameter caissons sunk 28 m deep for a pipejacking contract in New York in 1992. [127] Further information on caisson design and sinking can be found in Tomlinson [92].

Segments, underpinned in the construction of shafts, are also used to build caissons which can be sunk. The bottom ring, with a slightly greater external diameter to provide clearance, is equipped with a cutting shoe which assists the sinking. At the surface, a concrete ring beam is cast around the caisson, but separate from it, to act as a guide collar during sinking.

Precast circular concrete sections sunk as a caisson now frequently provide economic and rapidly-constructed drive and reception shafts. These can be used in all types of soils and ground water conditions. The most common diameter is 2.7 m, although standard sizes include 2.0 m, 2.55 m, 2.7 m and 3.0 m. Ring sections up to 3.5 m are available. Larger sections, such as the 4000 to 5000 mm diameters used for shafts on medium-size pipejacking are usually cast at site.

The following description of the small precast caisson system developed by ARC in conjunction with Yorkshire Water [67] exemplifies the method.

A base section of the caisson, generally 1000 mm high, is provided with a steel cutting shoe which gives a nominal 10 mm overbreak. The base section is positioned in an initial excavation 1 m deep with a diameter some 400 mm greater than the internal diameter of the caisson section. The outer surface of the base section is covered with polystyrene sheet 10-15 mm thick. Before sinking, the polystyrene wrapping around the outside of the base unit is dissolved, using a suitable chemical to break the joint between the base section and ring beam. The annulus between the base section and the excavation is filled with concrete to form a ring beam. This provides both permanent ground support to the upper part of the excavation and a temporary guide collar for sinking the caisson shaft.

Subsequent sections are then added to the base section to give the required caisson height. A watertight seal between successive rings is achieved by the use of a neoprene sealing ring, which sits in a groove in the upper edge of each caisson unit. The base unit has a series of cast-in threaded sockets distributed around its perimeter. These receive threaded tie bars, which pass down through the ring sections and are tightened from the top section to put a compressive force on the seals.

To sink the caisson, excavation is by a hydraulic grab mounted on a suitable full-circle slew excavator. A radiused boring clamshell grab facilitates more efficient excavation. In very soft ground the caisson will sink under its own weight but in firmer ground it is necessary to use the arm of the hydraulic excavator to exert pressure on the top of the uppermost section. Any misalignment can be corrected by applying load to the high portion of the top section. Where the caisson is used as a jacking shaft, it is usually necessary to sink it at least 600 mm below the final manhole base. This depth may have to be increased in some soils and in areas of high water pressure. At the jacking level, it is usual to use 1.0 m deep sections, with the level of the horizontal joint between these two sections coinciding with the longitudinal axis of the pipe to be installed. This allows the thrust from the jack to be distributed over two units and, in the opposite face, reduces the cut-out for the pipe to one semi-circle in each unit so minimising weakening of each unit. In certain soil types, any annulus between the outside of the units and the ground may need grouting to ensure that jacking loads are transferred from the units into the ground.

In addition to considerations of required invert levels, the caisson depth, and particularly the depth of the concrete floor, must be consistent with stability for the prevailing ground conditions. For example in very soft soil, the base of the caisson must resist base heave induced by upward extrusion of soft soil. Where the caisson is sunk through ground water, a suitable concrete base must be cast, by tremie if necessary, before the caisson is pumped dry. Failure to do this may cause the caisson to sink or topple. For the working thrust or reception shaft, the full diameter units should extend at least 150 mm above ground level to prevent surface water run-off or debris entering.

To determine the maximum allowable depth to which the precast concrete ring units may be used, it is necessary to consider forces acting on the ring. Using the most severe loading condition (where the ground water table is coincident with ground level), the total force is compounded of two components — the active lateral earth pressure and the hydrostatic pressure. At any depth, z, these will exert a uniform radial pressure

$$p = z(k_a\gamma' + \gamma_w) \qquad \text{- eqn 11.1}$$

where k_a = Coefficient of active earth pressure
γ' = Submerged unit weight of the soil
γ_w = Unit weight of water.

The maximum pressure occurs when z = D, where D is the depth of the manhole. This pressure causes a compressive stress in the wall of the concrete ring equal to

$$[D(k_a\gamma' + \gamma_w)d] / [2t] \qquad \text{- eqn 11.2}$$

where d = the diameter of the ring unit
t = the wall thickness.

Clearly this value must be less than the allowable compressive stress for the concrete.

In calculating the jacking thrust which can be resisted by the soil, account must be taken of the fact that active and passive earth pressures will not increase linearly with depth but will attain a constant value beyond a critical depth, due to the soil arching around the circular shaft.

Advance stabilisation. The main issues of advance ground stabilisation reviewed in Chapter 10 apply equally to shaft construction. Dewatering is a common and relatively simple method of stabilising granular soils with a high water table. The water level is drawn down below the shaft invert by pumping. This can be done by continuous pumping from a well point system, which has a depth limit of about 6 m, or from deep wells, where greater depths can be reached. However, for difficult soils the options are more complex, sophisticated and specialised. The ground can be stabilised by pre-injecting chemical and cement grouts. In high water table areas containing silty soils too fine to be grouted, ground freezing is another, but very expensive option. This is specialist, site-specific work.

11.3 Safe entry and exit from shafts

Proper provision must be made for a shield to pass safely through (a) the wall of the drive shaft and enter the soil (entry), and at the end of the drive to pass through (b) the wall of the reception pit (exit). (Some commentators reverse the use of the terms *entry* and *exit*, which can be confusing [25]. In cohesive soils the provisions for entry and exit are straightforward except in the case of slurry shields: the concern is to preserve a safe pit structure and a correct alignment of shield as it passes out of the shaft to drive into the soil. When working under hydrostatic pressure or in soft clays, silts, sands and gravels, an additional need is to ensure that no soil and only minimum water flow back into the pits.

Outside the shaft, the soil adjacent to the points of entry and exit can be stabilised in various ways. Inside the shaft, some kind of seal is needed to close the annular space between the shield (and then the following pipes) and the holes cut in the walls of both drive and reception shafts. Sometimes a combination of soil stabilisation and seals are necessary. The seal arrangement will always be required for slurry shields in all types of soil to stop the backflow of slurry into the pits.

11.3.1 Stabilisation outside the shafts

In unstable soils or in the presence of hydrostatic pressure, the ground outside the shaft immediately adjacent to entry and exit points must be stabilised. In granular soils down to depths not greater than 6 m, dewatering may suffice. Non-cohesive soils can be stabilised by injecting cement or chemical grouts. In very difficult ground conditions where dewatering or grout injection would not be adequate, ground freezing is a specialist alternative.

Where stability is a greater concern than hydrostatic pressure, the area outside the shaft can be restrained by some kind of temporary structure. A line of sheet-steel piling is often constructed to form a secondary wall, which can be retracted as the shield approaches. Another type of support is a secondary wall of unreinforced low-strength concrete, which can be formed by bored piling. To exit the drive, the TBM is driven through this relatively weak structure. A third alternative is to form a protective arch outside the shaft using jet grouting or similar techniques.

11.3.2 Sealing eyes within the shaft

Any type of slurry shield requires a sealing ring to be inserted in the shaft wall from the inside to prevent pressured slurry blowing back into the pit. The seal itself is made of rubber or neoprene 20 mm thick and is mounted in a framework often called the *eye*, an example of which is shown in Figure 11.6. The seal is covered by a blind plate which is removed to allow the shield to enter the seal. The eye also incorporates a gate which holds back the inflow once the shield is in the seal and is then raised to allow the TBM and pipes through.

If the cutter head is in advance of the shield, care must be taken to ensure the seal is not damaged. Safe progress can be maintained by advancing the head without rotating the cutters. Exit eyes and entry eyes are similar. However, exit eyes may be of simpler construction but must have greater built-in tolerances to allow for any driving deviations. Figure 11.7 shows a typical exit eye. For cofferdam shafts, the eye is constructed from within the finished shaft, while for caisson shafts, the eye is precast in the lowest segment before the shaft is sunk.

Figure 11.6 Entry eye in jacking pit (photograph by courtesy of Hobas Durotec)

Figure 11.7 Exit eye at reception pit (photograph by courtesy of Euro-Iseki Ltd.)

11.3.3 Key issues for entry and exit of shafts

Four fundamental factors have to be taken into consideration:

(i) Size of pipe
(ii) Depth to invert
(iii) Soil and ground water levels
(iv) Adjacent structures

The size of pipe being driven can range from, say, 250 mm to 5000 mm. Entry and exit considerations can be quite different, even in the same soil conditions. The depth to invert can range from 1 to 100 m. Hydrostatic pressure and the behaviour of the soil described in detail in Chapter 9 must be allowed for in the design. If existing structures and pipelines lie close to the shafts, special care needs to be taken to ensure that soil is not lost. Other factors which will influence the arrangements for entry and exit include the size and shape of shafts, the type of TBM and the type of pipe. Taylor and Brown list the prime considerations in the design of exit and entry points to shafts [128].

11.4 Reaction walls

11.4.1 Reaction walls in shafts

The jacking load can be transferred onto the soil by means of reaction walls which can be of reinforced concrete, or in the form of steel plates or grillages. The load is first transferred on to the sheeting at the rear of the shaft and thence to the soil. Whatever its structure, the dimensions and stiffness of the reaction wall have to be so designed that the load is transferred onto the soil uniformly and within the allowable stress or displacement values.

Invariably the passive thrust developed by a thrust wall is calculated using the Rankine-Bell equation. For a wall of height H and a dry soil of unit weight γ, this leads to

$$T_p = 0.5\,K_p \gamma H^2 + 2c'\sqrt{K_{pc}} \qquad \text{- eqn 11.3}$$

where T_p = Passive thrust per metre width of wall
K_p = Coefficient of active earth pressure with respect to internal angle of shearing resistance.
K_{pc} = Coefficient of active earth pressure with respect to
c' = effective soil cohesion

In calculating the passive thrust, two vital considerations are often ignored. Firstly, for the Rankine-Bell equation to hold true, the wall must rotate inwards towards the soil as shown in the Figure 11.8.

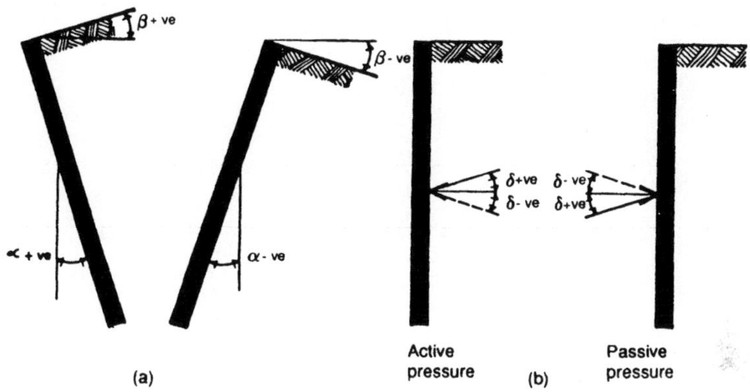

Figure 11.8 Wall rotation mode

Secondly, to mobilise the maximum passive thrust, T_p shown in the above equation, there must be movement at the base of the wall. This can be expressed as a fraction of the height of the wall and will vary according to the soil type behind the wall. The required movements for various soil types are set out below.

Soft clay	H/20
Stiff clay	H/50
Very stiff clay	H/70
Loose sand	H/20
Dense sand	H/100

As can be seen, these movements are very significant. For example, for a wall 5 m high they range from 50 mm to 250 mm. If the wall or the system cannot accommodate this movement, the full passive thrust cannot be developed. As can be seen in Figure 11.9, the relationship between movement and mobilisation of passive thrust is not linear. This means that quite substantial movement is required to mobilise even a small part of the passive thrust.

It should be noted that the analysis has assumed that (a) the full height of the shaft will act as a thrust wall and that (b) this will have smooth and intimate contact with the soil and (c) be free to move to generate full passive thrust. In reality the thrust wall is a block or grillage system bearing on the wall of the shaft. This wall is designed to support the surrounding soil and is carried out independently of the design of the thrust wall. Acting purely as an earth retaining structure, the bulkhead wall will have a complex regime of earth pressures acting on it. There may be active pressure acting on the back of the wall and passive pressure on the toe in front of the wall below shaft invert level. For multiple-braced shafts, an even more complicated earth pressure distribution pattern arises.

Unless the design of the bulkhead is executed to allow for thrust from jacking, the stability of both the thrust wall and the shaft bulkhead will be at risk. The jacking forces in an effort to generate a high passive thrust in the soil, will deform the shaft wall. This deformation can lead to the generation of large frictional forces on the rear of the bulkhead wall and a radical change in the simple earth distribution most probably assumed in the design. Where there is multi-framing this can result in dramatic changes in strut loads and yielding of struts can be expected.

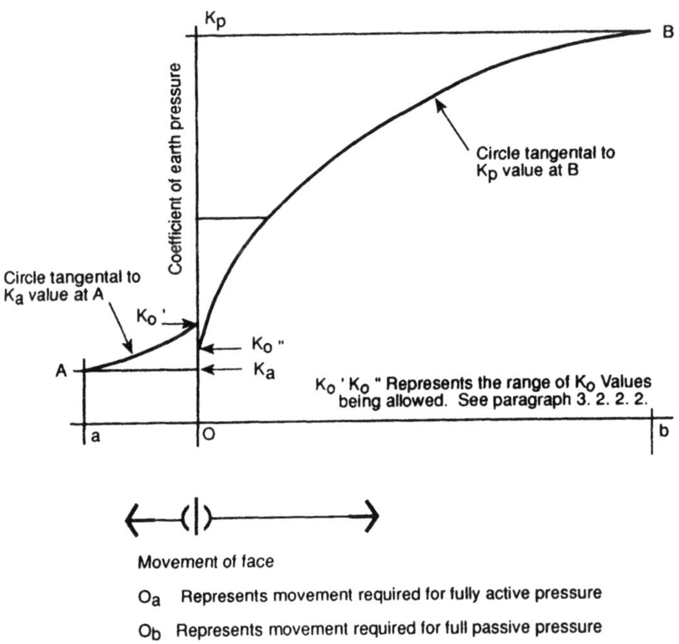

Figure 11.9 Relationship between movement and passive thrust mobilisation

The Swiss standard for jacking, SIA 195, [129] recognises that displacement can be the ruling factor in designing reaction walls. The graph at Figure 11.10 is based on data taken from many sites.

11.4.2 Reaction structures at ground level

Certain types of installation require the work to be carried out at or near ground level. Examples would be driving a subway or an underpass through an embankment. In such cases alternative methods have to be employed to provide the reaction to the jacking forces. The method selected will depend on the location, the total load and the ground conditions below the work.

A simple solution is to drive sheet-steel piles or bored piles with sufficient penetration to provide a cantilever type of reaction wall. In some soils, piles may need additional raked and compression piles to take the reaction. On a large underpass scheme at Brent Cross, outside London, the slab on which the underpass was cast was anchored into the ground below by a series of earth anchors. The jacks were then mounted on a carriage which ran on the slab. Heavy steel pins were used to lock the carriage into the slab and to transmit the load. [130]

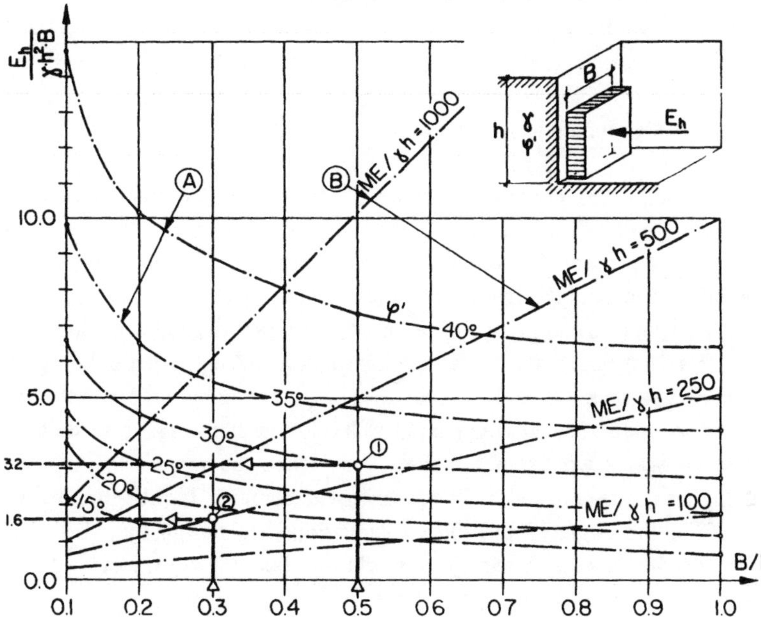

Where

E_h	=	safe jacking load kN
B	=	jacking wall width in metres
h	=	construction depth in metres
M_E	=	bearing capacity of soil behind wall kN/m²
γ	=	soil density kg/m³
δ	=	effective angle of soil friction

Curve A: bearing capacity is decisive

Curve B: soil deformation is decisive

Figure 11.10 Reaction wall design - SIA 95

Another approach is to *pull on your own boot straps*, a method used to jack a pedestrian subway through a rail embankment at Richmond, Surrey [131]. Sheet-steel piles were driven at entry and exit points. Horizontal tubes were bored through the embankment from entry to exit shafts. Steel rods with threads and couplings were then passed through the tubes to the exit side, extended through the working area and anchored to the sheet-steel pile abutment wall. The tension rods were fixed to a grillage at the rear of the jacking rig. The grillage became the reaction wall.

12 Design: contract documents

12.1 Responsibility for documentation

Contract documents are drawn up by the client, or his representative, in order that contractors can make their offers and the successful one carry out the work. The documents need to provide the definition and understanding of the work to be undertaken, how it should be done and the relationship between the parties.

Behind this simple description lie potential problems, misunderstanding, dispute and litigation. This is true for all forms of construction, but is particularly true for works below the earth's surface where the possibilities of encountering unforeseen conditions are much greater. An experienced engineer, Sir Harold Harding, once said wisely, "In most civil engineering works worthy of the name, the unexpected happens" [132].

The client has the responsibility for preparing the contract documentation, even though his own staff or the engineering group commissioned to design the works may in fact undertake to do this. In some parts of the world the client may turn to a specialist quantity surveying group to take on this responsibility.

Contract documents, and their contents, take many forms according to the nature of the work, the situation, the policy of the client, local custom and the legal framework. Normally they will comprise some combination of four sets of documents: Conditions of Contract, Bills of Quantities, Drawings and Specifications.

12.2 Conditions of Contract

12.2.1 Standard forms

The Conditions of Contract are normally based on some standard form that has been agreed by representative bodies for all parts of the industry. To these are attached the particular conditions which relate to the work to be carried out.

The standard form, such as the British Institution of Civil Engineers (ICE) Conditions of Contract [133] contain opportunities for conflict in nearly every clause. Contracts of this kind have resulted in the development of an adversarial industry.

Some large authorities issue their own conditions, while others amend and add clauses to standard forms. The objective in many cases has been to shift the risk for anything that happens away from the client and onto the contractor. There are many examples where clients have been disinclined to undertake site investigation or provide bidders with available geotechnical information lest claims are made if the information were proved to be wrong. This attitude has led to confrontational relationships between the client, his engineer and the contractor. The client has balked at the high cost of his works, which has become inflated because the contractor has had to cover his risks, and at the high cost of resolving any eventual dispute. Such contracts are often delayed because the parties' interests are not directed to what should be the common goal — the proper and timely execution of the work.

12.2.2 Time for completion

An area fraught with problems is determining what is a realistic period for completion of the works. The client may have very particular reasons for requiring the work to be completed by a certain date. For example, financial benefits may accrue to him if the scheme is put into operation early or he may have to meet commitments to third parties. The client's situation should be made clear in the contract. Bonuses for early delivery or penalties for overrunning time should reflect the true benefit and burden to the client. In general, a contractor objective will be to complete the job in the shortest reasonable period, as by so doing he will minimise his costs. On tight schedules, the progress of microtunnelling and pipejacking can often be accelerated, albeit at some additional cost, by working a second and even a third shift and at weekends. A contractor may also introduce more equipment, including additional TBMs.

12.2.3 Limitations

In the preparation of contract documents it is important to review standard contracts and "boiler plate" clauses to ensure that they are compatible with pipejacking and microtunnelling requirements. Standard contracts were often framed to cover more traditional methods of working. For example, Clauses 11 and 12 of the ICE Conditions of Contract [133] make the contractor responsible for the interpretation of all soil and hydrological information provided by the client for the purposes of constructing the works and for any design responsibility the contractor has to assume. It is further envisaged that the contractor may have to make his own investigation in order to meet his obligations. For a pipejacking or microtunnelling contract, such a requirement could place an onerous and expensive burden on the contractor.

12.2.4 Allocation of risk

To meet the shortcomings of these traditional forms of contract, a number of procedures have been revised. These recognise that a proper allocation of risk can be truly cost-effective for the client, because under traditional contracts the contractor has to allow in his price for any risk. Also time and money are lost in dispute. These revised procedures can greatly improve client/contractor relationships and contain the time needed for completion.

The US Construction Industry Institute has stated [134] that: *"The ideal contract - the one that will be most cost-effective — is the one that assigns each risk to the party that is best equipped to manage and minimise that risk, recognising the unique circumstances of the project."*

US experience. Constructability reviews have been introduced to try to ensure that what has been designed and specified can be practicably built. Value Engineering of design, a systematised form of review of methods and costs, is now common. Lack of experience and expertise amongst designers and specifiers of pipejacking and microtunnelling makes expert review procedures a worthwhile exercise.

Project Arbitrators or Dispute Review Boards are appointed prior to the start of the contract. Typically a three-man panel of experts will be appointed and charged with settling disputes as they arise. The experience has been that both client and contractor have accepted the decisions. Certainly they are less likely to enter into dispute if they know independent experts are arbitrating, rather than lawyers [135]. An additional procedure has been used on tunnelling contracts, which is said to have reduced risk and eliminated legal bills. As well as regular contract documentation, client and contractor agree to a three-point plan:

(i) The client guarantees the geotechnical report and so takes on the risk.
(ii) The contractor prepares an Escrow report, giving a breakdown of the components of the bid. This is kept in a bank vault.
(iii) Both parties agree to a disputes review board of three independent engineers, not lawyers.

If a dispute arises, the Review Board can then review all the documentation — a system which also deters parties from submitting ill-founded claims. Of the many US tunnelling contracts that have included this additional arrangement, none has so far led to litigation.

The perceived sub-surface conditions are at the root of risk allocation. To establish a geotechnical baseline the Geotechnical Design Summary Report (GDSR) [135] has been conceived in the USA. The GDSR should contain a description of subsurface conditions and ground behaviour. The report should also cover the following topics:

- Geological setting.
- Geological features of engineering and construction significance.
- Man made features of engineering and construction significance.
- Anticipated ground behaviour and construction difficulties.
- Influence on final design features.
- Influence on contract plans and specifications.

This information will greatly assist the contractor in preparing his bid and also in offering alternative solutions to match the conditions. If in practice the conditions are materially different from those anticipated in the Report, this situation will be more readily recognisable by all parties. The contractor would be entitled to compensation if, as a result of this discrepancy, he should suffer financially.

UK experience. Some of the privatised water authorities, including Thames Water and North West Water, have used the Institution of Chemical Engineers' contract form [136]. This cost-reimbursable type of contract with a target-cost formula is now widely referred to as The Green Book. This form of contract is felt to spread risk more fairly between owner and contractor and to allow work to start more quickly, often with an overlap between design and construction. The contractor is expected to participate actively in the investigation and design. This form of contract has been used for pipejacked and microtunnelled sewer installation.

12.3 Legal considerations

Contract documents must take into account and include specific reference to national, local and third party legislation and requirements. Existing requirements, particularly those relating to safety, the environment and disruption, have a substantial impact on the way the work is carried out and the cost of the scheme.

12.3.1 Safety legislation

Broadly, this legislation covers two categories of safety: public safety and the safety of workers.

Public safety. Although they may vary, legal requirements to protect the public during construction works have to be adhered to in all countries. Legislation to protect the public during works in the road normally include, but are not limited to, specific provision for the following:

Signing,	Illumination
Traffic control	Barriers
Hazardous material	Dangerous equipment
Prevention of unauthorised entry to sites	

Worker safety. Both pipejacking and microtunnelling are inherently safer for workers than the alternatives of open trenching and tunnelling. However, potential hazards face operatives working from shafts and under the ground with mechanical and electrical equipment. Regulations exist in all countries to minimise risk and to ensure safer site working. In his paper, Neale reviews the British Health and Safety Executive regulations and the impact of EC regulations on construction sites [137].

Particularly stringent are the regulations relating to the safety of men working in compressed air. No design and contract documentation for this type of work should be prepared without full account being taken of the relevant national and local legislation or codes of practice. In the UK, the so called Blackpool Tables form the basis for such regulations. In the USA, the Washington State Standards have been widely adopted as a model and in Germany the 1972 Compressed Air Ordinance.

Normally the rules become progressively stricter and the periods of working shorter as pressures increase beyond 1 bar. In pipejacking, however, rarely do pressures exceed 1 bar, as pressure balance shields will offer an alternative. A valuable reference to compressed air working is the CIRIA Report 44 [138].

Site safety regulations throughout the world are being strengthened to reduce the high level of death and injury associated with construction works. It is worth noting some examples of the legislation affecting microtunnelling and pipejacking.

USA

The Occupational Safety and Health Administration (OSHA) has greatly tightened up requirements on trench and excavation works — a likely spur for trenchless techniques. However, the OSHA is also expected to watch out for poor practice in some trenchless methods. Poor shoring to drive pits has been targeted as a problem area. The OSHA also regulates the conditions in which workers operate in confined spaces. These restrictions impact jacking work.

UK

In the UK, BS 6164 Safety in Tunnelling, issued in 1990, concerns pipejacking methods and makes reference to pressure balance machines. British Standards are advisory documents which become mandatory when included in contract requirements. BS 6164, *Safety in tunnelling*, specifies that the limit for man-entry tunnels be 900 mm, and emphasises the safety hazards of working in smaller diameters and the need for pre-planning and trials for rescue arrangements. The document also refers to ventilation and sets out formal procedures for checking for any deterioration in air quality or the presence of toxic gas.

Germany

The general rules for working underground in Germany are contained in UVV publications, *Construction work* and *Safety regulations for construction work underground* [139]. Safety requirements relevant to jacked work are also in a document *The safety rules for pipeline construction work,* VBG 37 [140]. This states that workers can only work in pipe if the diameter exceeds 800 mm. For pipejacking lengths more than 5 m the minimum diameter is 1000 mm. An amendment is proposed to increase the diameter to 1200 mm for jacking distances over 100 m. Directives are also issued dealing with health and safety issues for working in contaminated soils.

Europe

The harmonisation of European standards has been the wowrk of the CEN (European Standardisation Committee). CEN standards, to be mandatory for the 18 participating countries (EC plus EFTA nations), will have a major influence on equipment, methods and materials, including those for trenchless installations. These standards are being prepared under the direction of a group comprised of representatives from national standards bodies. These Regulations, which will take precedence over national regulations of members of the EC and EFTA, are scheduled to come into force in 1993. Safety standards for four categories of tunnelling equipment, including pipejacking, are under Technical Committee 151, Working Group 4.

Standards refer to:

- Hard rock machines
- Soft ground machines
- Road headers/impact rippers
- Compressed-air equipment

Standards will include recommendations on minimum sizes for man-entry working. A proposed guide line is for a minimum clear rectangular opening of 750 mm by 450 mm. This is equivalent to a 900 mm pipe. (Some participants are recommending that a minimum man-entry size should be 1200 mm for drive lengths in excess of 80 m).

12.3.2 Environmental legislation

Disposal of polluted soils and water. Polluted soils and their disposal have become urgent issues in many countries. In the USA, regulations laid down by the Environmental Protection Agency (EPA) have curtailed the range of materials that may be dumped in municipal landfills. At the same time these regulations have required more spoil materials to be taken to official secure landfills. This has discouraged the use of trenchless methods which create slurry wastes.

However, laws governing working in and the disposal of polluted soils have also brought benefits to pipejacking and microtunnelling. Jacking methods can operate with closed spoil systems and minimise exposure by workers to these materials. They generate a much smaller volume of spoil than traditional trenching, so the costs of disposal are less.

A parallel situation arises where the discharge of polluted water into water courses and storm water drainage systems is regulated by law. Any trenching work which involves pumping ground water polluted by leakages or spillages of fuel or chemicals must comply with regulations requiring that water to be treated before discharge. In high water table areas, well-point dewatering is often necessary before pipes can be laid by open-cut methods. As large volumes of water can be generated in this way, the expense of its treatment can add greatly to the cost of the job. Pumping and treating polluted ground water can be avoided by using counter-balance slurry microtunnellers for such sewer installation.

Ground water lowering. A second environmental concern with ground water is the possible damage to trees and shrubs caused by ground water lowering for construction work. Some cities in Germany, such as Hamburg and Bremen, have local laws restricting the use of such methods.

Noise. Most countries have regulations on permitted maximum noise levels, including noise emanating from construction sites. With reasonable attention to detail it is not onerous to provide equipment and barriers to keep within prescribed levels. If continuous working is planned, it may well be necessary to implement more stringent measures, including completely enclosing the works in a temporary housing. Other measures will be to limit trucks for soil disposal and pipe deliveries to normal working hours.

Disruption. Legislation to minimise disruption to the community will continue to proliferate, although many laws are at local and regional authority level. The rules for working in the streets of Tokyo for example are extremely stringent. In some congested areas the period of working may only be a few hours between midnight and 6 a.m. All shafts and trenches must be plated over each morning to allow traffic to use the area during the rest of the day [36]. Both Hong Kong and Singapore have legislation which restricts or even bars construction in certain roads [141] [142].

Regulations concerning both safety and disruption will severely limit the time and manner in which work can be carried out at crossings. Reference has already been made to the numerous codes and rules laid down by authorities responsible for railways, highways and rivers. It is important that both their general requirements such as Codes of Practice and their specific conditions for the job are carefully stated in contract documents.

12.4 Specification

Specifications are prepared for the purpose of describing the standards for workmanship and materials required on the contract. Specifications cover the technical details as well as general items required for a particular project.

Most countries have a great many recommended practices which are issued by professional organisations and standards associations. Some of the best known examples of bodies that issue standards and codes of practice are the American Society of Civil Engineers (ASCE), American Society of Testing and Materials (ASTM), the British Standards Institution (BSI) and the German Deutsches Institut for Normung (DIN). The international body is the International Standards Organisation (ISO).

For the most part, these standards and codes do not constitute statutory requirements. However, as they are recognised throughout the industry, they are adopted by clients and become part of the binding contract.

Limitations. Unfortunately, because of the period required for their preparation and publication, standards and codes trail technical developments. For instance, although in the USA concrete pipe has been jacked for some 70 years, no ASTM standard exists for such pipes, nor is one likely to be available before 1995. A second limitation is that many codes and specifications firmly embodied in contract documentation were based on conventional forms of construction. In an endeavour to limit their professional liability, many engineers are not prepared to consider a construction method unless they can use a recognised code or standard. These time lags in formulating standards and codes have inhibited the wider use of pipejacking and microtunnelling. This is particularly noticeable in the USA, where contractors are more likely to submit bids which conform closely to existing specifications than to offer alternative proposals.

Performance specifications. One way of overcoming difficulties caused by specifying a method is to specify what is to be achieved, the finished product and performance. Performance specifications are appropriate where specialist methods or equipment are proposed. Such an approach is suitable for pipejacking or microtunnelling bids, as the end product and its performance can be specified relatively easily. However, the contractor must have sufficient expertise to submit a detailed statement on how the performance is to be achieved. The engineer needs to have sufficient knowledge to evaluate the statement.

Not surprisingly, specifications for pipejacking and microtunnelling have been developed furthest in the countries which are leaders in this technology — Japan and Germany.

Japan. Both for the materials used and for workmanship, standards and specifications for trenchless methods of installation are well established in Japan. The Japanese Sewage Works Agency (JSWA) issued their first standard for pipes in 1973 [143]. The National Hume Pipe Association also produces standards for concrete pipes.

Any new technique is obliged to go through a period of testing and trials before standards are codified. Without this kind of recognition and its implicit approval, authorities will not employ new methods. Such control has both positive and negative effects. On the positive side, every new method is independently tested and trialed. Once accepted and specified, its use by authorities will be routine. On the negative side, the process is expensive and a new product takes a long time to come on the market. Also there is little incentive to improve or change a method once it is established with its relevant specifications.

Germany. The final version of ATV-Standard [144] for the structural design of pipe for jacking and for quality assurance to ensure greater safety in driving was issued in January 1990. It concerns the calculation of stresses acting upon pipes through the different stages of jacking and how to deliver proof of safety, this being based on the concept of probability of failure. It is interesting to note that the safety coefficient in direction of the pipe axis may be reduced to 80% if forces and eccentricity are properly analysed and calculated and the work executed according to the guidelines given. This standard also covers the structure and working method of quality control systems for drains and sewers as they apply to microtunnelling methods of construction. It aims to ensure maximum levels of safety when new drains and sewers are installed.

A German standard, *Regulations for Pipe Jacking Operations*, [145] became effective in 1992. These regulations apply to:

- Underground installation of precast pipe sections of any profile by jacking, driving or pulling.
- The cutting of earth and rock during pipe-jacking operations, as well as material transport from the pipe and immediate excavation area.
- Pipe-jacking operations using the displacement method.
- Replacement and displacement of existing lines.

The construction of drive and reception shafts is specified in DIN 4124.

UK. In the UK, the standard for concrete jacking pipes of man-entry diameters is BS 5911 published in 1989. It contains a number of important requirements including a test for joint face strength and a tolerance on squareness of ends. A proposed addendum to BS 5911 covering non-man-entry pipes (DN 150-900) was published in draft form for comment in 1991. This work is now *on hold* as the CEN drafting of a European Standard takes precedence.

Reference has previously been made to the work on the use of microtunnelling for sewer installation sponsored by Yorkshire Water Authority (YWA). Experience gained from this successful programme has prompted YWA to prepare a Code of Practice for Microtunnelling, which has been widely distributed [67].

This Code provides information on:

- Available methods of microtunnelling and their performance.
- Guidance on design with specifications for pipes and manholes.
- Contract documentation.

CEN. The new CEN Standards which have a bearing on pipejacking and microtunnelling materials and methods come under CEN 165 which is concerned with *Waste Water Engineering.* A series of working groups (WGs) are presently drawing up standards and specifications for the following:

- Production standards for the manufacture of vitrified clay and concrete pipes.
- The harmonising of the structural design and laying of buried pipelines - JWG 1, *Structural design of buried pipelines*, a joint working group with CEN/TC 164 (water supply).
- Sewer installation - WG 10
- Planning of works - WG 22

12.5 Drawings

Drawings, which show the proposed work in greater detail, should be read in conjunction with the specification. Together they provide a fuller picture of the work to be undertaken, the materials to be used and where and how they are to be installed. In the absence of any detailed Bill of Quantities, drawings define the work.

Drawings for sewer installations are more useful if plans and sections are kept together on the same sheet. If information about boreholes and ground water is added to the sections, a better evaluation can be made of the conditions in which the work is to take place. It is also important to include existing underground structures, pipelines and cables in the vicinity of the construction.

12.6 Bills of Quantities

Bills of Quantities are by no means universal and where they are used they range widely in the detail they provide. In British practice, these documents are highly detailed and have created a great deal of work for quantity surveyors, for those who prepare the bid documents and for those who work with them as a basis for payments. Typically some agreed standard method such as the *Civil Engineering Standard Method of Measurement* will be used as the basis for their preparation.

The stated objectives of Bills of Quantities are:

- To provide the contractor with an accurate picture and a basis for tendering.
- To enable to compare tenders on an equal basis.
- To serve as a basis for evaluating the work for payment purposes.

When provided, Bills of Quantities should be read in conjunction with all drawings and specifications. On their own, a set of estimates of the volume of materials will provide no picture or even a basis for tendering for a tunnelling, pipejacking or microtunnelling bid. Any contractor, therefore, who develops his estimate on unit rates multiplied by the quantities is likely to get into difficulties. Most contractors in this field estimate on a resource and time basis.

In the USA and other parts of the world which do not have the benefit of quantity surveyors, the client may not provide any quantities at all. Sometimes only a relatively brief list of the major items, such as total volume of excavation and concrete, will be given. Where this is so, the drawings and specifications must be in their final and fully detailed form as they provide the definition of the work. A dubious advantage of Bills of Quantities is that they provide a means for clients and engineers to make changes to the work during the contract and yet retain a basis for measurement and payment without amending the contract.

12.7 Submissions

Prequalification. Some form of preselection of bidders is now commonplace. For larger contracts, authorities wish to ensure that only contractors with appropriate experience plus adequate financial and management resources bid for work. To decide which contractors who claim to have jacking experience should prequalify to bid for pipejacking and microtunnelling contracts, clients must know exactly what kind of *jacking experience* they have. This umbrella description can be misleading. One contractor may have expertise with pressure balance TBMs for installing sewers in difficult ground. Another's experience may be limited to installing casings in cohesive clays using an open shield.

Bidders are therefore often required to submit details of how they propose to carry out the work. If a performance specification is submitted, this is essential. Such details will also be called for when an alternative bid is offered.

The following check list itemises areas that need to be taken into consideration by the client and his engineer. This list may not be fully comprehensive for all situations, and for others it may be over-elaborate. However, it is offered as a guideline.

CHECK LIST FOR PIPEJACKING AND MICROTUNNELLING

1. Equipment

1.1 Shield/TBM

Classification
Dimensions
Method of excavation
Torque
Face control
Pressure range control
Jacking system
Access for maintenance & breakdown
Procedures for breakdown
Obstruction procedures
Unforeseen ground procedures

1.2 Monitoring & steering

Monitoring method
Frequency of monitoring
Information to be monitored
Information to be recorded
Limits of deviation detection
Target position
Camera & data transmission
Steering system
Deviation correction

1.3 Material transport

System proposed
Breakdown procedures
Maintenance procedures
System limitations
Fluid formulations

1.4 Jacking arrangements

Estimated jacking loads
Jacking provisions in drive shaft
Jacking provisions at IJSs
Positioning of IJSs
IJS closure

1.5 Load reduction

Method
Coatings
Fluid formulation & quantities
Layout of injection system
Injection methods & pressures

2. Shafts & topside

2.1 Shafts

Shaft locations
Shaft dimensions
Advance works
Construction method
Design of temporary structure
Ground water pumping
Reaction wall type & design
Entry & exit arrangements
Reinstatement

2.2 Spoil handling

Spoil handling method
Volume of material
Storage arrangements
Separation methods
Spillage

2.3 Top side

Equipment
Layout
Offices & accommodation
Fencings & hoardings
Traffic control
Pipe storage & handling

3. Materials

3.1 Pipes & joints

Pipe type
Service load design
Installation load design
Joint type
Sealing arrangements
Joint packers
Transport
Handling & storage

3.2 Permanent shafts

Type
Furnishing & completion

4. Safety

4.1 Third party safety

Authorised party safety
Risk identification
Accident prevention plan
Adjacent property
Existing pipes & cables

4.2 Worker safety

Authorised officer
Risk identification
Accident prevention plan
Rescue plans
Accessibility of face, line & shaft
Communication system
Dust control
Fire risk & control
Falling objects
Barriers

5. Environmental

5.1 Spoil disposal

Tipping location
Polluted spoil
Contaminated spoil
Spillage

5.2 Water

Disposal point
Polluted water
Contaminated water

6. Noise

Equipment noise ratings
Noise containment
Working periods
Truck movements

7. Programme

Time schedule
Construction schedule
Mobilisation of equipment
Preparation of shafts
Installation of equipment
Working cycle
Turnaround
Clean-up

8. Resource schedule

Permanent materials
Equipment
Site management
Supervision
Workers
Work to be subcontracted
Subcontractors

13 Site control and supervision

13.1 Site organisation and management

13.1.1 Site responsibilities

However carefully the pipejacking or microtunnelling project is designed, the materials specified and the contract drafted it can still fail if the implementation is not properly planned and controlled. The prime responsibility for successful installation rests with the contractor.

In traditional construction contracts the contractor's efforts are primarily directed towards time and cost control, while the client's representative directs his efforts towards quality control, adherence to the specification and good engineering practice. This traditional division of responsibility becomes less appropriate for jacked installations, where the contractor provides specialist equipment, expertise and experience. It is in the contractor's own interests to assume responsibility for planning and undertaking the work in a manner which meets the specification and ensures quality control.

Before any work is put in hand, the engineer should have approved the contractor's submission on the installation method he proposes to adopt and his Quality Assurance/Quality Control (QA/QC) procedures. It is important that the engineer should have satisfied himself that the contractor's proposals will meet the contract needs and his specifications in particular. It is unreasonable to expect that some site representative with limited experience and authority should be delegated to take this responsibility after the work has started. It is the duty of the employer's representative to ensure that the contractor's approved QC/QA procedures are being closely followed in addition to his normal duties of inspecting and testing the completed work.

Formalised QA/QC procedures have been established in Germany for microtunnelling and pipejacking installation by a Quality Assurance Association [146]. With the objective of building in contractor quality from the start, these procedures focus on four areas:

(i) Installation of the finished line to specification.
(ii) Use of reliable machines and methods.
(iii) Effective and efficient operational procedures.
(iv) Quality-conscious operatives.

Point (iv) recognises the importance of making site personnel fully aware of any hazards or areas of concern that are likely to exist so that they are able to adjust working practices and methods. For example, if the drive has to pass some distance below a vulnerable water main, they will know to increase the level of monitoring and to adjust counterbalance pressures. They will also arrange to get well clear of a problem area within the course of a single shift.

13.1.2 Operator training

Dedicated and experienced operatives who have mastered sophisticated equipment are critical to its successful use. In Japan, the Japanese Pipe Jacking Association (JPJA) has drawn up a scheme to train and qualify candidates as Pipe Jacking Managing Engineers.

In the UK, Bridgewater Technical College offers several training packages for microtunnelling and pipejacking operatives and engineers. These courses cover:

- Manpower selection, identifying aptitude.
- Appreciation of the fundamental engineering principles.
- Working knowledge of equipment including its capabilities, limitations and maintenance requirements.
- Operating skills with *hands on* experience through a simulator.
- Guidance on setting up equipment, assembling, breaking down and safety.

The microtunneller simulator, based on a Soltau RVS35, was developed in conjunction with Decon Engineering to provide users with direct experience of machine operation. The programme simulates the steering behaviour as it passes through a variety of ground conditions. Simulated control valves and a theodolite view of the steering head allow the operator to practise steering a straight course between jacking and reception shafts.

13.2 Site works: shafts and machines

13.2.1 Shafts

The basic decisions on the position of shafts, their method of construction and their use as drive or reception shafts should have been made well in advance of arriving on site. In addition, all necessary permissions from authorities should have been obtained and any utility diversions organised. Shaft construction is covered in Chapter 11 together with arrangements for exiting and entering pits. Sinking of the shaft will allow a most useful check on the soil investigation. For example, the larger excavation is much more likely to reveal the presence of boulders and cobbles than will be revealed by a borehole.

13.2.2 Machine considerations

The choice of the shield will have been made according to the ground information and the evaluation of the conditions as discussed in Chapters 7 and 9. A contractor may wish to use his own equipment, suitably modified, or if the work is of sufficient length, he may purchase an appropriate machine for the contract. Many manufacturers operate a contract hire scheme, often providing a competent operator. Despite the apparent high cost, this can be an effective option, particularly for relatively small contracts. The contractor needs to take into account a number of operational factors when choosing a machine, including:

- Vulnerability to soil changes
- Ease and rapidity of equipment assembly
- Ease of operation
- Performance and reliability
- Ease of maintenance
- Service and spares availability
- Machine costs
- Running costs

Depending on the information revealed during shaft sinking, it will still be possible to make a final choice on cutter heads, bits and slurry mixes.

13.2.3 Set-up efficiency

Although often given too little attention, the initial logistics of mobilisation, rig set-up and break out have a critical impact on costs, even on short drives. (Costs are discussed further in Section 14.3). On a 50 m drive, these activities can easily account for 20 to 30% of the job period. The time taken for this has varied by more than 100% between different types of equipment set up in the same circumstances.

13.2.4 Turnarounds

A *turnaround* is using the same shaft for drives made in different directions. To move equipment frequently from one shaft to another costs more than making turnarounds in the same shaft. Time and costs can be saved by careful advance planning to keep the number of moves to a minimum and to organise the sequence of activities. Contractors may therefore opt not to use all the permanent manhole positions as drive and reception pits, to save the downtime involved in frequently moving from one shaft to another. One approach is to drive a double length straight through the position of an intermediary shaft and to sink the intermediary shaft onto the pipe after completion of the drive. A preferable approach is to construct intermediate shafts in advance and then to drive straight through these to next shaft selected as a drive/reception pit. In this case the base sections of the shaft need to be prepared with *unreinforced eyes* and the chamber filled with sand to allow the shield and pipes to pass without a loss of pressure. After completion, the sand can be excavated and the top of the pipe cut to form a channel section.

13.2.5 Driving

Driving should aim for smooth, trouble-free production. It is better to aim at a steady average rate each day than to try to break any records. This way neither equipment nor operators will be put under undue stress and QA procedures will not be skimped.

The operator can control a range of machine functions. Some of these, such as slurry pressures, are fixed by ground and ground water conditions. The rate of flow in the slurry system must be sufficient to maintain minimum velocity to prevent settlement in the pipeline. Flow rates must also be matched to the volume of spoil material that has to be pumped. Integrated rig design, where the strokes of the jacks are matched to the length of the section driven, has ensured that the speed of jacking the pipe can easily match the rate of excavation and disposal. The operation of IJSs can add to the time cycle, but with proper circuit and control design of the hydraulics this is not normally a problem.

However, the rate of advance and the volume of soil excavated are dependent on the speed of jacking and the rotation of the head which cuts the soil. The operator needs skill and judgement to achieve the correct balance for varying soil conditions. By advancing too quickly there is a danger of stalling the cutting head. This relationship between torque at the head and speed of pushing has been studied by Piquereau [147]. In Figure 13.1 the power developed is shown as a function of the pressure applied. There is an optimal operational level which will maximise production without risk of stalling. Many manufacturers use electric drive motors in preference to hydraulic ones, claiming that temporary increases of up to 50% over rated capacity can be tolerated.

Data logging systems which provide a permanent record of all functions are recommended. They can provide valuable evidence of events and changes in conditions.

13.2.6 Adding new sections

Progress on most work is determined not by the speed of pipe installation into the ground but by the time required to add new sections to the line. This involves stopping other activities and the breaking of supply lines. The more sophisticated the operation (e.g. a pressure shield with slurry pumping), the more significant the adding of a new section becomes. Pressure has to be maintained in the line during this operation and a pit bypass flow unit is required to isolate the slurry flow. Contractors can minimise downtime by preparing in advance the new lengths complete with all temporary pipework and power lines. It is not uncommon to see a length of pipe driven in 10 to 15 minutes whereas anything up to 50 minutes can be taken to add the next length of pipe and recommence driving. Undoubtedly then the progress rate will depend greatly on the experience and dexterity of the operatives adding pipe lengths. Mistakenly, some contractors tend to concentrate their attention on the machine operator and use relatively inexperienced workers for the operations in the shaft. Where space is available and work is at shallow depths, the use of longer shafts and maximum pipe lengths will increase productivity.

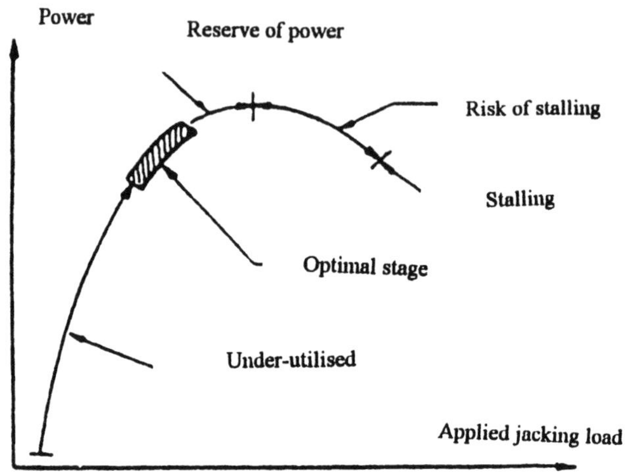

Figure 13.1 Relationship between torque and speed of advance (graph after Piquereau, 1989)

If the pipeline has to be tested after completion of the drive, this should be done before the jacking rig is removed. Should a leaking joint or damaged pipe be found it will be a relatively inexpensive operation to jack in new pipes until the defective section emerges.

13.2.7 Obstructions

A significant number of cases has been reported where pipejacking and microtunnelling work has been brought to a stop because a remote-control TBM has run into an obstruction. A troublesome condition is a rock horizon which rises through the invert level. This may cause the shield to ride up the rock surface rather than to cut through it.

Chapters 3 and 5 have already commented on machines equipped with crushing capability, but the size of obstruction that can be taken into the shield is limited to 20 to 30% of the diameter. Where the boulder or obstruction is large, it has to be broken down by the cutter head. Several machines are fitted with special roller bits or cutting discs to break up the obstruction at the face. Dynamic tools using *down-the-hole* hammers have also been used in machines like the Japanese Super Striker.

The greatest difficulties occur where a large obstruction is encountered in a matrix of softer material. The size may be too great for a crushing machine and cutting tools are often not effective because the obstruction will move in the softer material rather than be cut. Several manufacturers now claim that with the combination of the correct head and skilled operatives it is possible to bore through boulders. In Switzerland, local contractors have developed equipment to deal with the rocky Swiss soil conditions, including a facility for coping with large boulders [148].

Should an insurmountable obstruction be met, it is possible to bore a vertical shaft mechanically and to break up or even to remove the obstruction, provided there is surface access. Such an exercise would be costly and time-consuming. Undoubtedly the more prudent approach would be to conduct better site investigation and/or use line-proving methods in an attempt to identify large obstructions at the design stage. If work is to take place in ground known to contain significant obstructions, it may be necessary to consider alternatives to remote-control closed face machines. In stable ground, an open-face shield can be equipped with a boom cutter to provide the flexibility needed to deal with a variety of conditions. In ground which is both unstable and likely to contain obstructions, compressed air working may be a solution.

13.2.8 Working in rock

Although there are now available machines and cutting heads capable of cutting rock, close attention needs to be paid to the nature of the material and its behaviour when broken down. Reports from a variety of jobs have shown that debris can get forced or jammed into the annular overcut and create very high jacking forces. A similar problem has been observed in sandstone or mudstone where the abraded particles are of fine sand and silt size. They fall and get deposited under the shield and pipe and, when water is present, turn to a slurry. In either case the pipe tends to ride up and jam against the top of the annular cut. This situation was reported on a contract in Nuremberg. In another case, fine particles mixed with the material being used as a lubricant set around the pipe. To overcome such difficulties a number of counter measures have been devised. Bentonite slurries, with a designed gel strength to keep particles in suspension, are injected to keep the annular void open. On a German pipejacking contract, a foam filler was injected immediately behind the shield to fill the space and stop material entering the gap [149]. The foam is produced at the injection point using a resin and foaming agent.

13.3 Monitoring and control

13.3.1 Controlling line and level

The importance of closely controlling the line and level cannot be overemphasised. Firstly, there must be close control of line and level to comply with the specification. Secondly, such control, by keeping the deviations within close tolerances, makes the operation more efficient as there is a substantial reduction in jacking loads. Longer lengths can be driven without introducing an IJS and damage to pipes and joints by overstressing is eliminated. Small adjustments should be made early by responding even to small deviations and in a manner that ensures a slow smooth return to the line.

Many monitoring systems can predict the anticipated line and level after, say, a further length of pipe is installed. This valuable facility lessens the tendency to overshoot when correcting.

13.3.2 Angular deviations

Line and level tolerances are commonly quoted in terms of maximum allowable deviations. Perhaps a more important criterion in terms of severity of stress concentrations is the angular deviation at the joints. Instrumented field trials [109] have clearly shown that peak radial and shear stress coincide with maximum angular deviation. In Section 10.2.7 this was discussed in detail. The French Society of Trenchless Technology have proposed a test requiring a pipe to accept a misalignment of 1° under maximum design thrust force [49]. Other authorities have suggested that the maximum angular deviation allowable should be 0.5°. However for most soils, with due care it should be possible to substantially reduce the maximum angular deflection to 0.2° or less. In very weak soils which provide poor support this will be difficult to achieve.

There has long been a belief that deviations will progressively be smoothed out as jacking proceeds thereby improving the alignment. There is no evidence to support this convenient concept.

At the start of a drive the procedures to maintain line and level control need particular care. Launching a TBM into the soil can be difficult and time consuming. The transition from the machine being rigidly supported in the shaft and held by the wall as it passes through into softer material often causes settlement of the head or the whole shield to drop when it is clear. Misalignment at the start of a drive is particularly undesirable in that all the pipes have to pass through these deviations. As the drive length increases, the jacking load reaches its peak so that the final pipe sections are the most heavily loaded and also subject to maximum stress from deviations. It is this combination that often leads to pipe failure.

It is important to note that angular displacement needs to be monitored under load. Most surveys of pipe jacked lines coincide with break times when the load is not being applied and angular deflections relax.

13.3.3 Steering systems

The criterion for a good steering system is that it gives accurate information about the current position and attitude of the shield. Some more sophisticated remote systems are now capable of detecting deviations of 0.1 mm. Although fully automatic artificial intelligence programs are available which analyse the data and generate corrective steering commands, many contractors still prefer the flexibility and experience of a skilled operator to control the steering and other adjustments. Current specifications often require closer control of level, typically ±25 mm or ±35 mm. Line is less critical and specifications quote ±50 mm or ± 75 mm. In consistent ground conditions it should be possible to stay easily within these tolerances.

For pipejacking, the conventional theodolite survey is still quite widely used, particularly for non-straight drives in cases where the cost of a fully automatic laser system is not warranted. However, among mechanised monitoring methods, laser-based systems predominate. Many are used with open shields. A laser system has the advantage of providing a constant reference. The ability to ensure a continuous supply of reliable information is fundamental for this type of equipment. A number of problems or shortcomings have already been identified and these are addressed in Chapter 3. The main points of concern are:

- Movement of the laser in the drive area. This can occur due to malicious or accidental interference. A more common problem is movement of the shaft under jacking loads. Since this often happens imperceptibly as the jacking load increases, it often goes undetected.

- Dispersion and distortion of the laser beam seems to be a particular problem in smaller machines where heat and atmospheric pollutants are being generated.

- Opinions amongst manufacturers differ about the effective working distance of laser equipment. The effective range quoted seems to vary between 150 and 250 m. The use of active targets with diodes can provide a computerised analysis of the centre of the beam even when diffused. This arrangement will extend the working range of a laser.

- Keeping the beam on the target is sometimes difficult when the machine is deviated. This may mean no further line and level information is generated and the operator has the option of driving blind or stopping work.

13.4 Safety

13.4.1 Man-entry diameters

As has been discussed already, minimum man-entry diameter, although by no means universally standard, is generally put at 900 mm. For smaller diameters all driving operations are carried out remotely and maintenance and repair operations have to be undertaken at a shaft. For pipejacking, the length of drive is an obvious factor that also has to be taken into account in any safety consideration.

Although for most countries legislation on safe diameters and lengths does not yet exist, it is likely to be introduced shortly. Common sense indicates that, both on safety and economic grounds, DN 900 mm pipes and drive lengths of 60 m should be a maximum. For larger pipes, say, DN 1200 mm, with open shields it will be technically possible with IJSs to drive over virtually any length. The economics of restricted working space and spoil disposal coupled with safety considerations will in fact present a practical limit of around 200 m.

It is a commonly held view that the use of remote-control machines avoids the problems discussed above. This is not so, as man entry to the rear of the shield may be required for maintenance and repair. Working space may be restricted by spoil disposal lines and pumps and this needs to be taken into account in rescue planning. In 1993, OSHA introduced safety regulations for work in confined spaces, which could greatly inhibit man-entry jacking in smaller diameters in the USA [150].

13.4.2 Explosion

Some sites may contain pockets of methane, a by-product of organic decomposition in landfills or previous swamplands and marshes — a well known hazard in tunnelling. Strict regulations exist for checking for the presence of methane, its monitoring and the working methods to be adopted where it is suspected. Such regulations apply equally to pipejacking and microtunnelling. Modified spark-free electrical systems will lessen the danger of underground explosion in the presence of volatile gases.

13.4.3 Ventilation

Most countries have laws stipulating an adequate level of ventilation for any tunnel construction. Monitoring the exhaust to detect any contaminants is now commonly required. This is additional to the gas monitoring for combustibles and oxygen. Ventilation systems may have to deal safely with fumes, dusts, pollutants and any risk of poisonous or volatile gases. Such systems will be required in open shield pipejacking for the protection of workers. With remote-control machines, ventilation for workers will be required where men have to enter the tunnel for maintenance or repair work. It may also be necessary to provide forced ventilation for cooling equipment and reducing laser distortion.

13.4.4 Lighting and communications

A good lighting system which complies with the requirements for tunnel working is needed at the face, along the line and in the shaft bottom. More general flood lighting is required around the shaft top. An efficient communication system linking the operatives in the working areas with each other will not only assist safe working, but also greatly facilitate normal operations.

13.4.5 Shaft area

Good housekeeping around the top of the shaft and proper slinging arrangements for lowering pipes, tools and materials will help prevent injury from falling objects, perhaps the most common form of accident in civil engineering. Safety helmets should be compulsory for all operatives.

The arrangements for shafts accessed by ladders should be in accordance with safety regulations. Ladders, rails and barriers all have to be securely fixed and landing stages will be required for deeper shafts. As tunnelling often involves mud and water, a hazard on ladders and landings, these areas should be kept clean.

13.4.6 Use of chemicals and additives

Chemical grouts are often used to stabilise the ground through which a bore is to be driven or to strengthen the shaft areas. Various polymers are used in slurries and lubricants. Most proprietary grouts and additives have been tested and approved to ensure they are safe to handle and compatible with the environment. Manufacturers' instructions on handling and storage should be carefully observed. Their use may be restricted by environmental concerns in certain soil and ground water conditions.

13.4.7 Hazardous spoil

Operatives are at risk in areas polluted by domestic or industrial waste. Other high risk sites are at airfields or near fuel storage tanks where discharges have accumulated in the soil. Spoil from excavations may be classified as hazardous or toxic and require special handling and disposal at licensed sites.

Remote-control machines offer one of the safest and most convenient methods of working in such conditions. Operatives have very limited contact with the material at the separation and disposal stages; their protection is relatively straightforward. Where men are working in an open shield, it has to be recognised that the confined space and limited access make such an operation potentially hazardous. In the USA, OSHA safety regulations for working with hazardous waste include operations in tunnelling [151]. These include 40 hours of health and safety training on toxicology, chemical and physical hazards emergency procedures, use of protective clothing, respiratory equipment, medical monitoring, air monitoring and first aid. To implement these measures, carrying cumbersome respiratory equipment and wearing protective clothing for instance, is difficult in the confined space of a tunnel. Mobility can be restricted to the extent that workers may be impeded from operating effectively or even safely. Periods of work should be limited for men using a respirator.

14 Economic considerations

14.1 Variability of costs

This chapter considers costs under a number of headings:

- Costs for project appraisal
- Cost of estimating for pipejacking and microtunnelling
- Key cost variables for pipejacking and microtunnelling.
- Typical current cost levels - direct and indirect
- Costs of crossings
- A comparison with costs of open-cut construction

Costs will depend on ground conditions, volume of work, type and size of installation and other variables. Costs will also vary according to geographical location; marked differences in prices are noted from one country to another for the same type of work. Within the same country there will be also significant differences. For example, the cost of undertaking the same type of pipejacking installation in New York or Boston can be twice the cost of the same job in Dallas or Houston. When indirect costs of disruption to the community are included, total costs for installing a sewer in a heavily trafficked main street may be several times the costs for the same installation in a quiet side street in the same city. The nature of civil engineering work is such that each job has to be considered as a unique undertaking with its own specific problems, solutions and costs. This comment is particularly true when work is below the surface in developed areas of cities and towns.

14.2 Project cost appraisal

Project appraisals aim to assess both the costs and benefits of a proposed scheme before selecting from all the alternatives the one which will bring the greatest benefit at the lowest cost. This approach is generally accepted as a basis for evaluating a range of capital works. However, if it is to be realistic, a project appraisal must consider the total costs of alternative forms of installation. Where utility works come into conflict with the public interest, these total costs must include social costs (the costs to the community) as well as the direct and indirect costs of the project.

Direct costs include:

- Planning, design and supervision.
- Payments to contractors and suppliers.
- Diversion of existing utilities.
- Permanent reinstatement.
- Traffic diversion and signing costs.

Indirect costs include:

- Compensation for damage to land and property.
- Compensation for damage to utility plant.
- Compensation to businesses for loss of profit.
- Reduction in pavement life.
- Increased road maintenance.

Social costs include:

- Disruption and delay to traffic.
- Increased accident levels.
- Loss of public and private amenity.
- Environmental impacts, e.g. noise, vibration.

Generally speaking, the utility pays for direct costs. However, direct costs such as for reinstatement or traffic diversion are often overlooked in the project appraisal. On the principle that the disturber pays, the utility organisation should be liable for indirect costs, such as those listed above. In practice, these are not always included in the calculation, and are frequently disputed. Therefore the utility does not always pay. If indirect costs are not paid by the utility, they become a charge on the community and add to the social costs.

Until recently the utility has never paid for the social costs. However, the situation is changing and legislation, both at national and local level, is being introduced that in one way or another makes the utility at least take these social costs into account.

14.2.1 Direct costs

In assessing direct costs the following general points should be noted under the headings listed above.

Planning, design and supervision. The planning, design and investigatory work for pipejacking and microtunnelling as discussed in Chapters 6 and 7 may be more extensive than that normally employed for open-cut solution. However, for any type of sewer project the engineering cost represents only a small percentage of the total value, therefore an increase of 10 or 20% in the cost of planning, design and supervision will have little significance in the assessment.

Payments to contractors and suppliers. The cost of the work is expressed in terms of the expected final payments to be made to all contractors and suppliers, not on the basis of bid prices. There is often a substantial gap between the two.

The first requirement for a realistic assessment of the proposed costs of the work is that like is compared with like. At first sight, a comparison of direct costs proposed for a job seems relatively straightforward, but this is not always the case. Experience indicates that cost figures provided by clients, engineers and contractors can be made up in a variety of ways.

Diversion of existing utilities. When the work is under urban roads, diverting existing utilities can be a major cost component of an open-trench project. Yorkshire Water Plc. allow 15% of the value of the work to cover such diversions [152]. Some diversion, although much reduced, may still be required at the shafts for microtunnelling and pipejacking. However along the line, the costs for diverting buried utilities are eliminated.

Permanent reinstatement. The demands of highway authorities concerning the amount and standard of permanent reinstatement have increased markedly during the last few years. (see Section 14.2.4) Open trenching of a deep sewer down a road can cause damage to road surfaces well beyond the nominal trench width. Some authorities are now requiring that the whole width of carriageway be finally surfaced. Pipejacking and microtunnelling methods cause little disruption and therefore the costs of reinstatement are minimal.

Traffic diversion and signing costs. The associated direct costs for diversionary signing, police control, advertising etc. are a direct charge that must be taken into account in the project appraisal. The need for traffic diversion and attendant precautions is again greatly reduced for trenchless installations [153].

14.2.2 Indirect costs.

Evaluating indirect costs can be more complex. In the past, *indirect costs* have not been properly recognised as separate cost items and therefore have not been allowed for in cost assessments. Costs legitimately chargeable to the utility carrying out the work have either been at rates less than the real costs or have not been charged to it at all. These indirect costs, listed above, can represent a substantial part of the total cost of the work and are worth further comment.

Compensation for damage to property. Damage to land or property caused by installation work can be immediate or delayed. Immediate damage, for example to a householders garden, can be easily identified and assessed. More important is the delayed or long-term damage that may be caused to structures. As an engineer does not knowingly design an installation project which may cause settlement to adjacent buildings or other structures, he is unlikely to allow much, if anything, in his cost assessment for repair or compensation against this item. In reality there is a high incidence of settlement to property and structures as a result of trenching and tunnelling works. As settlement will often take place over an extended period, claims may not be agreed for years after the work has been completed.

Compensation for damage to utility plant. Similarly, damage to existing buried utilities can be immediate or delayed. More difficult to allow for are future remedial costs for damage to pipes or cables caused by long-term settlement of a trench. These are indirect costs nonetheless.

Compensation to business for loss of profit. In recent years, a utility company's liability to compensate businesses for loss of profit because of damage, cut services or restricted access caused by disruption has become an indirect cost of greater significance. As a result of some court rulings, utility companies can no longer disclaim liability for such costs as easily as they once did. Although attitudes are changing, the situation is often unclear on liability. In the UK for example, compensation claims for loss of trade can be made against the water utilities undertaking sewerage work but not against other utilities. Such are the difficulties and costs of bringing a case, that legitimate claims are not always pursued. The present level of compensation paid for loss of business is substantially below the value of real losses incurred.

Reduced pavement life. From the results of studies undertaken in the USA by the Public Works Association (PWA) and the Gas Research Institute (GRI) and elsewhere, it has become evident that cutting trenches into a highway shortens the pavement life. The period of time by which it is shortened is still a matter of some debate. Findings to date suggest that pavement life is reduced by some 15-30%, although obviously the pavement construction, ground conditions and nature of the trenching work will influence the percentage by which its life is reduced. If this percentage reduction in pavement life can be determined, the cost of this can be calculated fairly simply.

The cost of additional maintenance work increases after trenching. Temporary resurfacing works are usually charged to the utility. However, even after permanent reinstatement, maintenance cost in the area of trenches is likely to be higher than in areas not cut by trenches. Once permanently reinstated, the cost of further maintenance is normally carried by the highway authority — a substantial indirect cost to that authority. Such costs are difficult to assess, but in most cases are ultimately borne by the community as *social costs*.

14.2.3 Social costs

The privatisation of many utility companies, increased environmental concern and new laws that reflect this have helped establish the relatively new concept of *social costs*. It is now accepted that these costs to the community do exist and that they should be allowed for when an installation project is assessed. Recognised social costs of the type discussed here, although they will never be easy to quantify, are certainly going to influence the installation policies of utilities and contractors in future.

Loss of amenity. To open-cut a deep sewer across a public park may mean loss of trees, shrubs and flower beds or deprive the public of an amenity because it has temporarily to be closed. Although the social cost may be neither large nor easily quantifiable, public reaction is likely to be negative. As utilities become more sensitive to such reactions, they will be more likely to choose a different line or a less-disruptive installation method.

In environmentally sensitive areas, not only the wrath of the community will have to be taken into account but also the growing volume of environmental legislation. There is now virtual prohibition of any form of surface construction in sensitive areas like wetlands and natural forests.

Noise and vibration. Vibration, noise, dirt and air pollution are all part of traditional civil engineering construction. Both noise levels and vibrations are measurable and their impact can be assessed. Certainly installation methods which generate lower noise and vibration levels will be more acceptable.

Level of road accidents. A report from the British Transport Research Laboratory has provided evidence that construction works on roads lead to an increased incidence of accidents [154]. It is suggested that the cost to the community in deaths and injury is a cost that should be included in any assessment. There are established methods of costings for accidents used in appraisal calculation.

Traffic disruption. Of all social costs, the most significant and the most hotly debated is the cost of traffic delays. At present, the parameters affecting traffic flow and how it may be impeded by utility works have not been precisely defined, but they include:

- Volume of traffic along the road.
- Location of the works in relation to junctions and bends.
- Extent of the obstruction to the traffic lanes.
- Length of carriageway affected.
- Alternative routes.
- Delays on diversion routes.
- Access for vehicles and pedestrians in the area of the works.

It has been estimated that the cost of disruption and delay to traffic caused by trenched sewer installation can be several times the value of the direct cost of the work. Studies of urban utility construction sites [38] have shown ratios of social to direct cost varying between 0.1 and 10.5. In one particular study the mean value for 14 sites was a ratio of 3. A more recent study of 10 sites [155] showed ratios varying between 0.5 and 5, with a mean value of 1.3. Of all types of utility construction, sewer installation is the most disruptive and occupies the street for the longest period due to the more extensive nature of the work. Therefore it would not be unreasonable to assume a ratio of 2 for the indirect to direct costs for sewer works in urban roads. Pipejacking and microtunnelling methods do not eliminate all such social costs as there will still be some disruption, albeit greatly reduced, around shaft locations, but their level will be much lower than those arising from traditional trenching methods. If an even more modest ratio of 1: 1 for social to direct cost were accounted for in a project assessment, open-cut methods would be ruled out for the vast majority of sewer work in any urban and suburban situation.

14.2.4 Alternative approaches to social costs

Highway authorities and governments are growing more aware of the high level of indirect and social costs associated with trenching works that are being carried by the community. Since utilities have so far been generally reluctant to recognise these costs to the community, various measures have been introduced to ensure that disruption is at least minimised or that some payment is made.

Controlled utility working. To control the works of utilities is the most common approach, although the ways in which these controls are implemented differ. Generally work must be in accordance with specific restrictions, such as a requirement that it may be done at certain times only. In Tokyo, for example, permission for installation or maintenance is given as a *concession* by the road authorities. In the city centre, open excavation is permitted only between midnight and 06.00 hours and all excavations have to be decked over for traffic use during the day. In the suburbs excavations can only be made between the morning and evening rush hours and are not permitted during busy times [36]. In Hong Kong, excavation permits are issued with severe restrictions on times and methods of working [141].

Time prohibition. The New York Authority gives notice to all utilities of its intention to repave a section of street in Manhattan. During this time utilities have the opportunity to carry out any new installation or rehabilitation works. Once the Authority's work has been completed, utilities are denied access for a period of several years.

Cost penalties. Some authorities deter utilities from opening a trench by imposing a penalty. There may be no prohibition as such, but a deterring requirement, for example, that the whole length of the road affected should be totally resurfaced afterwards at the utility's expense.

Road space charging. The British New Street Works Act of 1991 [37] introduces a new concept of charging for road space occupied during either construction or maintenance work. The charge would depend on the importance of the highway and the amount of traffic. As the bill passed through Parliament, opponents of this concept prevailed on the government to make provision for these charges as a reserve power only. The understanding is that these charges would be introduced at a later date if they were found to be necessary.

Road rental charges. Road rental is not the same as *road space charging*. The idea of road rental is to minimise the time taken by contractors on work. Normally the contract specifies the maximum period of occupation. Should the time be exceeded, the contractor must pay a penalty. Conversely, bonuses can be won for early completion. Versions of this system, modified to cover utility work, have been used in several countries.

14.3 Estimating costs

The basic approach to estimating the cost of a pipejacking or microtunnelling project is resource scheduling, the resources being men, machines, materials and time. Accurate estimates require a good understanding of the technology selected for the project and its performance. The estimating process can be viewed as a series of stages.

Stage one is to define clearly the programme of work to be covered in the estimate. This definition should be according to the drawings, site investigation report and the specification. Any limitations, such as restricted site space, should be recorded.

Stage two, which relates directly to the first stage, is to determine the basic methodology by which the work is to be carried out. This includes the choice of a machine which matches the geotechnical requirements, the shaft construction, and pipe type and size. Space availability and the location of the work will obviously influence the assumptions made.

Stage three is to prepare a programme and allocate resources to the main activities. Some of the resources needed, such as the quantity of pipe and number and size of shaft rings will be straightforward. Others, like cranage and generators, will be determined by technical needs, such as the loads to be handled, or by limitations of site layout. The availability and relative cost of alternatives will come into the decision-making. Allocation of men in terms of skills and numbers will be based on experience and past performance on similar projects. The allocation of time requires an element of judgement, again based on experience and on the evaluation of the work as defined at the first stage.

Stage four is to prepare the estimate and arrive at the direct cost of the job. To this must be added site overheads and insurance, usually expressed as a percentage of the direct costs, the total giving the site cost. Head office overheads and profit margins for contractors, also expressed as percentages, are added to arrive at the total cost of the work. In addition there will be contingency sums, or risk factors, due to unknowns arising from the work itself or to the extent of the estimator's knowledge. A clear statement should show what each of these sums is for and the amount allocated for each item. In the estimate itself realistic numbers should be used for time, men, machines and material. The temptation to add contingency or ignorance factors at this stage should be avoided.

Examples of estimates for a specific microtunnelling installation, the job plan, the allocation of time for each procedure and estimated costs per item are shown in Figures 14.1, 14.2 and 14.3, taken from a paper by Moss [156]. It is interesting to note that of the 18 days allocated for the installation programme, 5 days (28%) were for set up and turnaround. The microtunneller used was a Decon RVS 250A and shafts and manholes were ARC caisson systems. Figures 14.4 and 14.5 show the relative importance of the site costs. The first pie chart at Figure 14.4 shows the work broken down into activities. The shafts represent 35% of the cost (44% with the finishing works), the pipes 17% and the actual cost of installation 39.5%. Figure 14.5 divides the costs of the work into its main components — materials, labour and equipment. Leaving aside the finishing works to the manholes, of the remaining site costs of £54,810, 54% is for material, 33% for plant and only 13% for labour. As all estimates must be job and site specific, clearly these percentages will vary for differing types of work, but they provide some guidance on the key cost elements in a microtunnelling project.

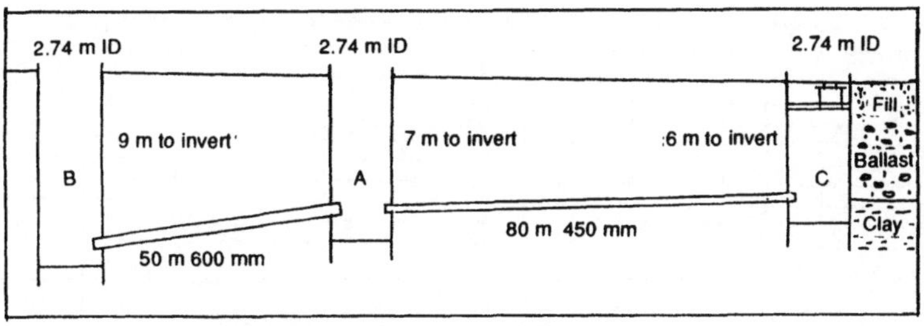

Figure 14.1 Cross-section of a microtunnelling job (after A. Moss [156])

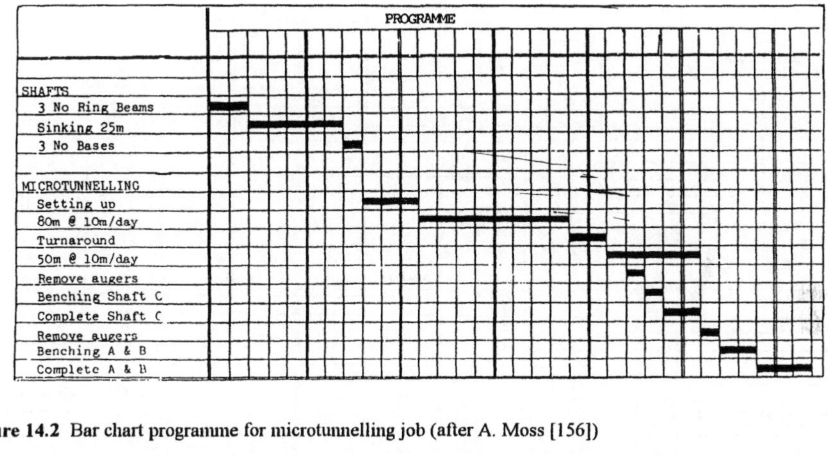

Figure 14.2 Bar chart programme for microtunnelling job (after A. Moss [156])

14.4 Key cost factors

This section reviews the individual elements of the cost estimate and in particular the key determinants of performance, pipes, shafts and installation as shown in Figure 14.4.

14.4.1 Pipe

The price of pipe of different types is set by suppliers and will show some variation between countries. The cost of pipe for a particular installation depends on its length and the type specified. As already discussed in Chapter 8, pipes for jacking must be of additional strength and quality. This requirement plus the cost of special joints makes them considerably more expensive than standard pipes for trenching. Jacking pipes can represent anything from 15% to 40% of the site cost. Table 14.1 shows the relative costs of different types of pipe in the UK. Table 14.2 provides a range of pipe costs in the UK in 1993.

ESTIMATE

Description	Quantity		Unit Rate L	Unit Rate P	Unit Rate M	Labour £	Labour £	Plant £	Plant £	Material £	Material £	£
SHAFTS												
Foreman	8	D	70			560						
Labourer	8	"	45			360						
Excavator driver	8	"	55			440	1360					1360
20t excavator	8	"		100				800				
Tools & tackle	8	"		30				240				
Compressor	8	"		20				160	1200			1200
2.7m ID ring	25	m			500					12500		
2.7m Cutting edge	3	No			1500					4500		
Concrete	15	cm			50					750		
Muck away					600					600	18350	18350
										TOTAL SHAFTS		20910
MICROTUNNELLING												
Operator	20	D	110			2200						
Fitter	20	"	80			1600						
Labourer	20	"	45			900						
Fork lift operator	20	"	55			1100	5800					5800
RVS 250A	20	"		700				14000				
175 KVA Generator	20	"		80				1600				
Fork lift	20	"		30				600				
Tools & Tackle	20	"		40				800	17000			17000
600mm pipes	50	m			90					4500		
450mm pipes	80	m			70					5600		
Muck away					500					500		
Craneage between shafts					500					500	11100	11100
										TOTAL M/TUNNELLING		33900
MANHOLE CONSTRUCTION												
Foreman/finisher	9	D	70			630						
Labourer	9	"	45			405	1035					1035
Compressor	9	"		30				270				
Formwork	9	"		15				135				
Reinstatement equipment	9	"		20				180				
Tools	9	"		30				270	855			855
Concrete	10	cm			50					500		
Ladders	17	m			35					595		
Reducing slabs	3	No			500					1500		
Access shaft/cover	3	No			150					450		
Reinstatement	3	No			120					360		
Craneage					300					300	3705	3705
										TOTAL MANHOLES		5595

SUMMARY

Shafts	20,910
Microtunnelling	33,900
Manholes	5,595
	60,405
Add site overheads & insurances @ 12%	7,249
Total site costs	67,654
Add Head Office overheads @ 8%	5,412
Total estimated cost	73,066
Add Profit @ 15%	10,960
	£84,026

i.e. 130m @ £646

Figure 14.3 Estimate for a microtunnelling job (after A. Moss [156])

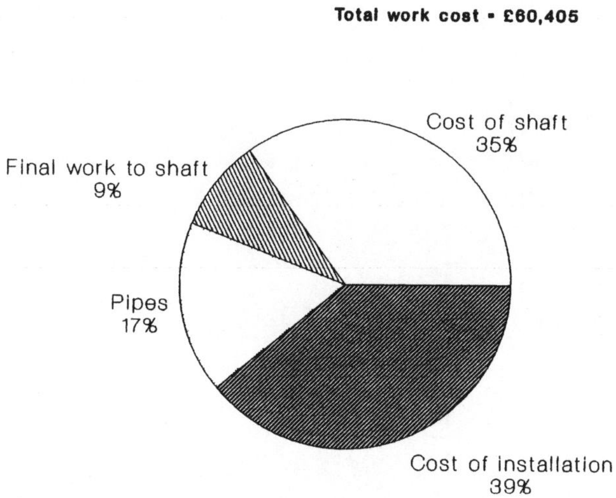

Figure 14.4 Percentage breakdown of activities on a microtunnelling job (after Moss [156])

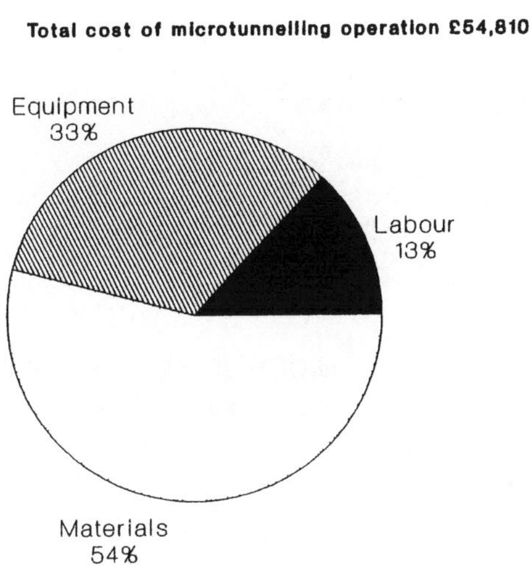

Figure14.5 Main cost components of a microtunnelling job (after Moss [156])

Table 14. 1 United Kingdom: Relative pipe costs (DN 450 mm)
[100 units = standard length concrete with steel collars]

Concrete	100
VCP (rubber joint)	180
GRP	140
Ductile Iron	240
Steel	200

Table 14.2 Range of pipe costs UK, 1993

Pipe dia (mm)	Type	Unit cost £/m
Microtunnelling		
300	Concrete	50 - 65
300	Clay	80 - 90
500	Concrete	70 - 90
500	Clay	140 - 150
500	GRP	110
600	Concrete	95 - 115
600	Clay	165-185
600	GRP	150
Pipejacking		
900	Concrete	130 - 160
900	GRP	270
1200	Concrete	170 - 190
1200	GRP	360
1500	Concrete	205 - 240
1800	Concrete	300 - 325
2100	Concrete	380 - 420

Individual pipe lengths. The choice of individual pipe lengths is largely a pay-off between the benefits of reduced downtime during pipe positioning and the need for increased shaft length to receive longer pipe sections. Pipejacking presents fewer problems for adding new sections so the downside of shorter length pipe is less important. In fact, the greater weight of longer sections can require larger cranage. A parametric study of microtunnelling reported that, for concrete pipe lengths, any increase over 1.2 m brings a negligible cost benefit [157]. However, this is not true for steel pipes as the downtime and cost of welding a joint is substantial.

Many clients and engineers have reservations about short length pipes, as they see the greater number of joints increasing the possibility of leakage.

14.4.2 Shafts and pits

Costs of shafts and pits depend on their size and depth, ground conditions, reinstatement and site access. Clearly it is cheap to install shallow pits in good ground in a greenfield site and extremely expensive to install deep pits in bad ground in a congested urban site.

For shafts over about 3 m in depth, the trend is to sink a caisson. Some indicative rates for this type of construction are given later in Table 14.3 (Section 14.4, Budget costs). As shaft diameters increase beyond 3 m, costs can rise dramatically, particularly where ground stabilisation treatment is needed. For shafts of 5 m diameter, costs of £4000 per m depth have been reported, even in good soil conditions.

Shaft depth. The additional unit cost variation on the total installed cost of constructing deeper working shafts and permanent manholes is not a major factor in the installed cost. For example the additional cost of deepening a microtunnelling shaft by 3 m will add about £2500 to the cost. When this is spread over a 100 m drive length the add-on cost to the installed price is £25 per metre. Such an increase is a small fraction of the cost of going down an extra 3m by open trenching methods. Further reference is made to the influence of shaft costs for crossings in Section 14.6.

14.4.3 Installation

Whether pipejacking with an open shield or a pressure balance remote-control microtunneller is being considered, the cost of installation can be broken down into four main areas:

(i) Mobilisation costs
(ii) Set-up costs
(iii) Operational costs
(iv) Machine costs

The main concern, as always, is to reduce the cost of the resources and time expended. Sometimes both can be reduced, but frequently it is a question of balancing the pay-offs.

(i) Mobilisation costs
Naturally mobilisation costs will vary with the size and type of equipment. The costs of mobilising all the equipment required for a pressure balance slurry microtunneller will be much greater than those for an auger machine. Mobilising a 2000 mm full slurry TBM will be several times that of mobilising a semi-mechanised open shield. The essence is the relationship between the mobilisation cost of equipment and the volume of work to be undertaken. In the case of one-off crossings, this cost will be very significant. However, for a sizeable length of installation, it will be less significant in relation to total value of the works. On a contract involving 1000 m of installation, a mobilisation cost of £10,000 is only £10 per metre; on a crossing of 100 m, it is £100 per metre.

(ii) Costs of set up and turnaround time

For microtunnelling, the time taken to set up the equipment in the drive shaft, turn around to drive in the opposite direction or to move equipment from shaft to shaft will have a significant impact on the cost per installed metre. For example, on a scheme where a series of drives average 80 m, the typical driving time for each length may be 5 shifts, of which an average 3 shifts may be required for setting up and moving. Driving a crossing of 40 or 50 m may take only 2 or 3 days, in which case the set up time may exceed driving time. As drives get longer for pipejacked installations, the impact of set-up time on the cost per metre diminishes.

Once again it becomes evident how important careful advance planning can hone costs. In Chapter 13 it was shown how driving through intermediary shafts can reduce the downtime for set-up and thereby the cost per installed metre.

(iii) Operational costs

Cost per metre of pipe installed will be a function of the all-in cost per shift (labour and plant) and the production achieved in that shift. These items are considered under the following heads:

- Operational cost of the equipment (reviewed in the next section).
- Average drive rate.
- Size and cost of the labour force.

Drive rate. The important factor is the full cycle of driving operations, not just the rate of machine advance. The average drive rate for *pilot* and *two-stage* systems of installation are relatively slow because of the additional activities involved. Although two-stage systems can have technical benefits, for the average run they will be hard-pressed to compete with one-stage systems on cost. Elvidge [157] has shown that increasing the rate of driving up to 100 mm/min can yield worthwhile benefits; any increase beyond this adds little additional cost benefit.

Labour costs. The size of the crew and its cost varies between types of equipment and even more so between countries. Obviously a large pipejacking installation will require a larger crew than a microtunnelling one. Typical crew size in Europe for microtunnelling and jacking up to 1500 mm will be 4 men: a machine operator, a top-side plant operator and a pit crew of two. At least double this number seem to be used in Japan. Union rules and safety regulations will also have a bearing on acceptable crew sizes.

Increasing the number of men is unlikely to increase production rates or quality of work significantly. Much more important is the efficiency and experience of the crew and how well they work as a team. Remuneration should therefore reflect their value. For most situations labour costs are almost a second order effect. The exception would be in the case of low capital cost equipment with a high utilisation level, for example, on crossings using open shield driving with manual excavation. Double rather than single shift working has relatively little influence on costs, mainly because the percentage of total time taken up by the driving operation is quite small.

(iv) Machine costs

Capital costs. Types of pipejacking and microtunnelling equipment range from relatively simple jacking rigs with an open shield to sophisticated remotely-controlled pressure balance shields with a separation plant. From one end of the range to the other capital costs can vary by a factor of 10 or more for a set of equipment to drive the same size of pipe. In general, the simple machine is written off over a longer period at a lower rate. The more sophisticated machine has a much higher capital cost and shorter life expectancy and consequently has to be written off at a much higher rate. However, the influence of the capital cost is essentially a straight line write-off which can be based on a period of time which might typically be 5 years. A more realistic base for estimating machine cost would be the hours worked or the length installed. At one time 5 km was the accepted point at which to write off a machine, but improved design and reliability have increased this length to 10 km or to a working period of 10,000 hours.

An alternative to purchasing is to hire a machine and skilled operator(s), a service offered by some manufacturers. This can offer a contractor an economic option, particularly for relatively short lengths of work, or if contracts are likely to demand differing solutions.

Utilisation rates. The results of Elvidge's comparative study [157] and others have all clearly shown that utilisation rate is the biggest single factor in reducing unit costs. Figure 14.6 demonstrates that for utilisation rates less than 500 m/yr, unit costs rise rapidly. However, unit costs are almost independent of utilisation rate once this reaches 1250 m/yr.

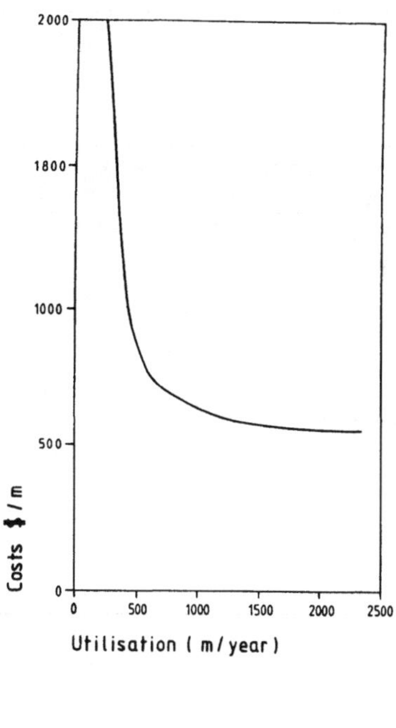

Figure 14.6 Effects of utilisation on machine costs (after Elvidge [157])

It has been noted that in Japan there are a great many machines in relation to the volume of installation undertaken and that the average installation rate per machine appears to be 400 m or less per year. These low utilisation rates (along with larger crews) must be a major factor in the relatively higher costs of installation in Japan than elsewhere. It should be added that the influence of tax legislation on depreciation write-offs can be a distorting element.

Until machines are more fully employed, relatively high unit costs will continue to counteract the benefits of more competitive machine prices. Characteristic market patterns are more encouraging. Unit costs for contractors are generally high initially, but then level off over two or three years as client confidence in jacking methods increases and rates of utilisation improve. Figure 14.7 shows this trend in unit costs for Singapore [142]. High initial costs per metre in 1983 fell by 40% by 1987 across all diameters and then stabilised over the following two years with relatively small increases. Yorkshire Water Plc undertook a series of microtunnelling and small pipejacking work [152] over the period 1986 to 1989. Figure 14.8 based on their experience clearly shows the drop in costs per metre over the period.

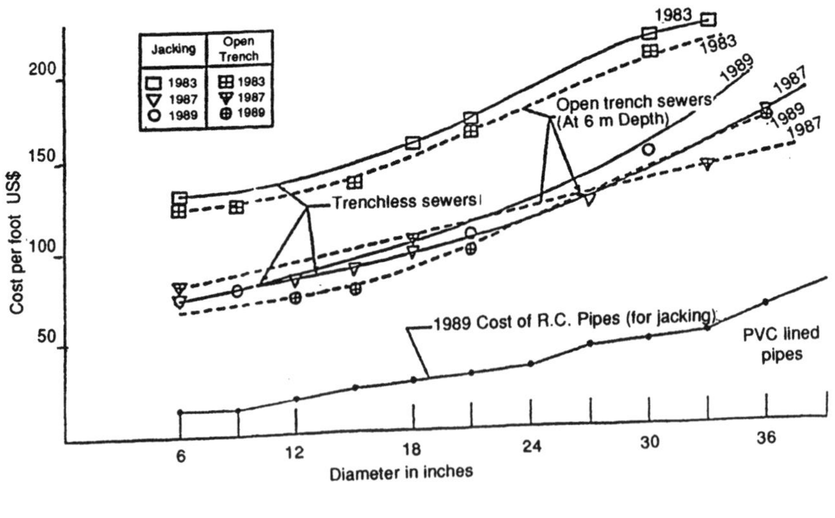

Figure 14.7 Costs of microtunnelling and open cut sewer installation (after Lim and Balasubramaniam [142])

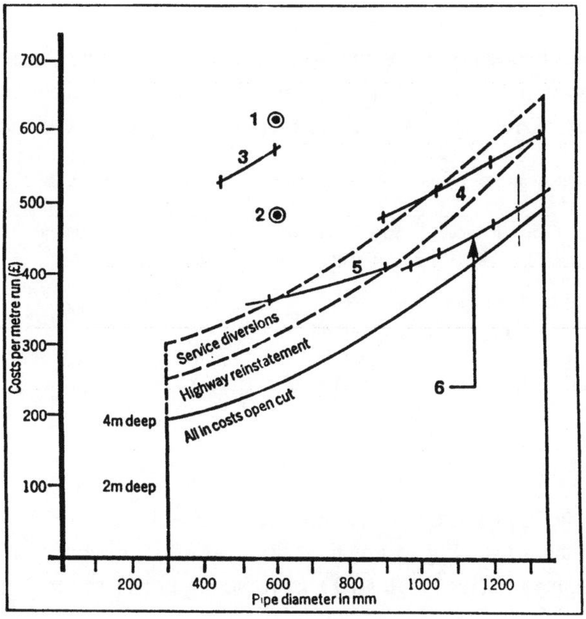

ABOVE RIGHT: Tender prices show how the cost of microtunnelling has closed with traditional methods in Yorkshire.

DIAGRAM KEY

Costs are for microtunnelling and pipe jacking on commercial contracts in Yorkshire Water Authority's North & East division, not work done during the 'WA/Decon/ARC Joint Venture. All costs have been standardised for comparison to 1989 prices.

Scheme A: Completed 1986/87

1 = Cost of microtunnel including enlarged working shafts.
2 = Cost of microtunnel excluding enlarged working shafts.

Scheme B: Completed 1987/88

3 = Microtunnel
4 = Pipe jack

Scheme C: Due to start September 1989, tendered Spring 1989.
5 = Microtunnel
6 = Pipe jack

Figure 14.8 Experience of microtunnelling and pipejacking prices 1986-89, Yorkshire Water Plc. [152])

Maintenance and running costs. Maintenance costs can be related to the capital cost of the machine and its anticipated service life. Some contractors take the cost of maintenance as 50% of the capital cost spread over the 10 km service life. Consequently, for a £400,000 machine the unit cost for maintenance would be £20/m. Equipment running cost can be calculated from past records or some allowance can be made, which, depending on the machine, could be £10-£20/m. Local differences and fluctuations in fuel prices will also have to be included in the estimated cost.

14.5 Budget costs

Frequently, it is necessary to calculate approximate costs either to estimate cost levels or to compare these with costs of open cut. Table 14.3 shows some cost levels in the UK (1993) for some of the main elements for remote-control pipejacking and microtunnelling.

Table 14.3 Cost elements for remote-control pipejacking and microtunnelling in the UK (1992)

Pipe size (mm)	Concrete pipe £/m	Installed cost £/m	Shaft cost £/m	Mobilisation £	Moves / turnarounds £
300	60	170	700	7 000	2 000
500	80	240	700	10 000	2 000
600	105	270	1 200	10 000	3 000
900	145	300	1 200	10 000	3 000
1200	180	380	1 500	12 000	4 000

By choosing the appropriate combination from the table it is possible to obtain an idea of the total installed cost. For example, 1000 m of 300 mm concrete sewer with shafts 7 m deep on average and drive lengths 100 m would require 11 shafts of a total depth of 77 m.

From the table we can calculate an installed price of £311 per metre, as

		£	£
Pipe rate	1000 m	60	60 000
Line rate	1000 m	170	170 000
Shaft rate	77 m	700	53 900
Mobilisation		7000	7 000
Moves/ turnarounds	10 (number)	2000 each	20 000
Total			310 900

Reference can also be made to patterns of past costs for similar work. Figure 14.9 shows a range of UK 1992 prices for microtunnelling with concrete pipe. It is worth emphasising that unit costs can be a useful guide in preparing estimates, either for traditional or trenchless works, but they cannot be substituted for a proper costing of resources.

14.6 Costs of crossings

14.6.1 Casings 900 mm or less

Where a steel casing of DN 200-900 mm or less is a requirement, installation by non-steerable auger boring is the most common method. Although microtunnelling has been developed for direct installation of the service pipe to close tolerances, this method is also used for installing casings. However, for most crossing work over short drive lengths this level of sophistication is not needed.

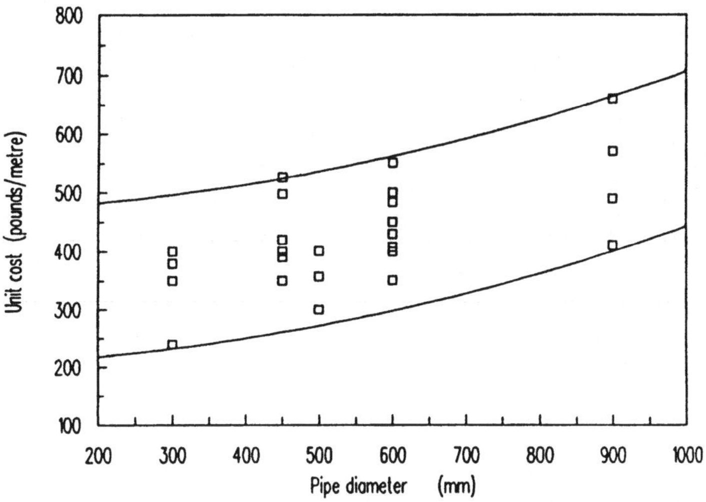

Figure 14.9 Range of all-in costs for microtunnelling - UK, 1992

Microtunnelling vs. auger boring. The installation of a steel casing by auger boring at a typical rail or road crossing will be carried out from shafts less than 3 m deep. Typical prices range from £210 to £275 per metre installed for DN 300 mm plain steel casing. For a 900 mm casing, rates range from £320 to £400 per metre installed. Thus a 50 m drive length will be installed for about £12,000 for a 300 mm sleeve and, say, £18,000 for a 900 mm sleeve.

The cost of installing a 300 mm pipe by microtunnelling at a one-off 40 m crossing with a 3 metre deep shaft at each end can be budget estimated from Table 14.3 to have a total price of £26,200. Thus an auger bore casing could be installed at an installed price of £360 per metre compared to the installed price of £524 per metre for microtunnelling.

Microtunnelling is only cost-effective at crossings where it has a technical, and thence an economic, advantage over auger boring. Such situations would include direct installation of a pipe, deep inverts, bad ground conditions, especially bad ground water conditions, or shallower crossings where ground disturbance must be kept to a minimum.

14.6.2 Pipejacking

Non-mechanised pipejacking, was originally developed for installing casings DN 900 and greater at crossings. Both steel and concrete casings can be used. The cheapest method is the simplest, with an open shield and hand excavation. Mobilisation and capital cost are low. This technique can be used successfully in good ground conditions where there are no ground water problems, hence shallow crossings are often suitable. For DN 900 mm pipe, output is good over drives up to 50 m long with all-in installed rates as low as £400 per metre. To undertake a short crossing using a pressure balance TBM would more than double this rate.

For more sophisticated full face TBMs to be considered, the length of crossing would have to be at least 200 m, the ground conditions difficult, or for there to be good geotechnical reason for anticipating ground disturbance.

Although very basic, the above examples for both microtunnelling and pipejacking illustrate that the mobilisation and shaft costs of different techniques combine with the installed length to have a profound influence on costs per installed metre.

14.7 Comparison with open cut

14.7.1 Traditional open-cut sewer installation

For open trench installations deeper than 1.5 m to 2.0 m, depending on ground conditions, costs start to increase rapidly because of the need for ground support and/or dewatering. As more hand working will be demanded, equipment becomes less productive. Even at these shallow depths, installation costs increase markedly when pipes exceed DN 300 mm.

Deep sewers with inverts from 4 m to 6 m in depth will incur considerable increases in construction costs for excavation and/or more substantial support systems. In bad ground conditions, support systems can easily add 30% to the installation costs of large diameter pipes at depths greater than 4 m. If the same type of work is in a developed area rather than on a greenfield site, costs may be several times greater. Additional costs will arise from a number of causes but in particular will be due to:

- The high percentage of hand excavation.
- The need to work around and protect existing services.
- The provision of temporary and permanent reinstatement.
- Restricted working space.
- The need to cope with pedestrians and traffic..

Typical proportional costs for sewer construction in a road are given in Figure 14.10. It is interesting to compare these costs with the breakdown of costs for microtunnelling in the earlier Figures 14.4 and 14.5. For the open-cut sewer installation, the value of the permanent structures, pipes and shafts, relative to total costs is much reduced. The greater part of the cost is represented by the temporary work of digging a trench and reinstating the road surface

Greenfield sites. The decision to switch from open-cut excavation to a pipejacking or microtunnelling solution on a greenfield site will essentially be made at that point where the increasing cost of trench excavation exceeds the cost of jacking. Reinstatement and social costs do not enter into the consideration. Whether or not pipejacking will be selected as a cost-effective method for greenfield sites will depend on:

- Availability of skilled contractors
- Soil conditions (e.g. pipejacking in soft clays may be easier and safer than trenching).
- Depth of the installation

Pipejacking becomes progressively more economic at greater working depths. In Germany depths of 7-8 m seem to be the economic break point. One reported jacked installation [158] on a greenfield site in the UK was for a sewer construction at depths between 7 m and 8.5 m.

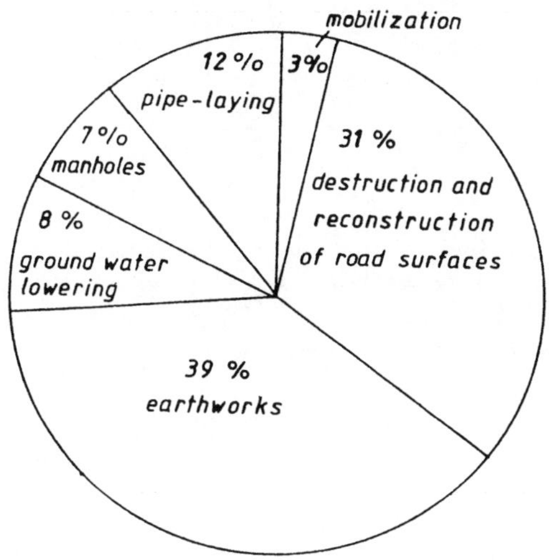

Figure 14.10 Cost analysis of sewer construction by traditional trenching, Berlin (after Knut Möhring)

Developed areas. When the installation is to be in a more developed location, the decision to select a trenchless solution becomes more complex. The costs of open-cut working become greater as other factors enter the picture. Yorkshire Water Plc have set out conditions where trenchless sewer installation should be considered, graphically presented in Figure 14.11. These conditions can be listed as:

- At depths greater than 4 m
- Under a busy road
- Under or near existing underground pipes and cables
- In unstable ground conditions
- Below the water table
- Where soil movement is a hazard
- Under sensitive environmental areas
- Under socially sensitive areas - gardens or parks
- In industrial and commercial areas

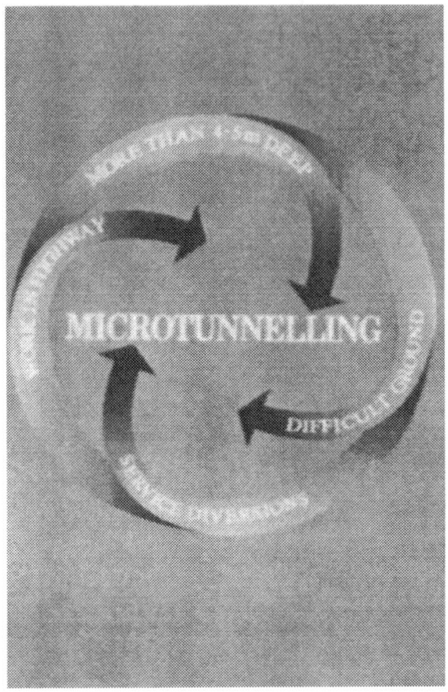

Figure 14.11 Conditions in which microtunnelling should be considered (chart from *Code of practice for Microtunnelling*, Yorkshire Water Plc)

If any two of these conditions prevail, a pipejacking or microtunnelling solution should be considered. Figure 14.8, already referred to and based on the Yorkshire Water contracts, shows that for work in roads at 4 m or deeper microtunnelling and pipejacking become competitive, particularly in sizes greater than DN 500 mm.

The authorities in Berlin are leading exponents of trenchless installation, with about 40% of their sewer installation work now done by microtunnelling. Since 1985 nearly 100 km have been installed. Figures 14.12 and 14.13 reflect this experience [159]. The depth at which microtunnelling becomes cost-effective is a little over 2 m, in this case working under concrete and asphalt pavements.

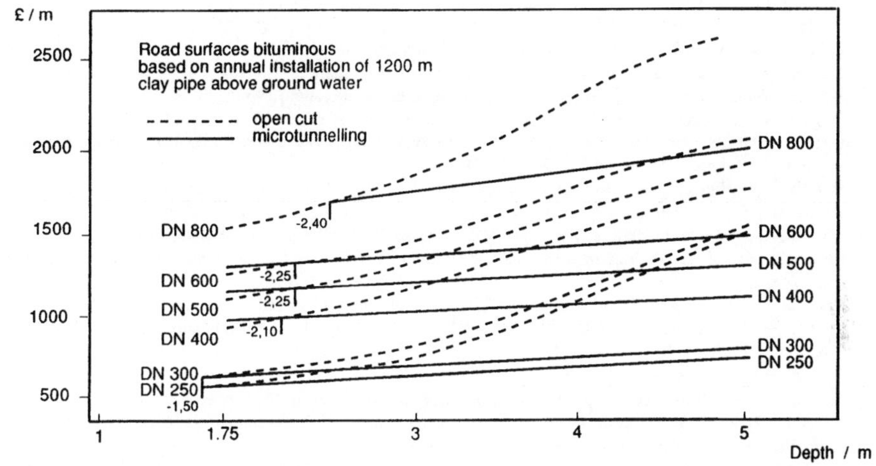

Figure 14.12 Comparison of microtunnelling with open cut in bituminous paved roads (after Möhring, Berlin)

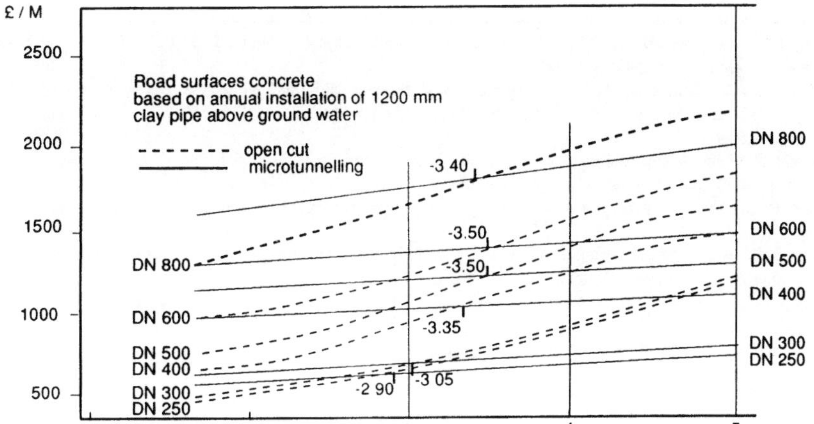

Figure 14.13 Comparison of microtunnelling with open cut in concrete roads (after Möhring, Berlin)

References

(1) Report, Railroad Gazette of the 10th Annual Convention, New England Road Masters Association, 1892.

(2) Loving, N.W. Concrete Pipe in American Sewerage Practice, *Bulletin 17*, American Concrete Pipe Association, 1938.

(3) Peckworth, H.F. *Concrete Pipe Handbook*, American Pipe Association, 1958.

(4) Griffin, A. Pipe culverts placed by jacking, *Engineering News Record*, May 1924.

(5) Feld, J. Jack-in-place pipe drainage *US Highway Research Board Bulletin*, 1948.

(6) *Concrete pipe lines, chapter IV*, American Concrete Pipe Association, Chicago, 1942.

(7) *Jacking reinforced concrete pipelines*, American Concrete Association, 1960.

(8) Die Technik der Betonrohrdurchlasse im Strassenbau, *Beton und Eisen*, 1934.

(9) Lenz und Muller Beispeile fur im Durchpressverfarhen eingbaute grosse Leitungen aus Stahlbeton und Spannbe tonrohren, *Beton und Stahlbeton*, 1970.

(10) Installing pipe-lines by the thrusting process, *Concrete and Constructional Engineering*, August 1964.

(11) Thomson, J.C. Horizontal earth boring, *Proc. Institution of Civil Engineers, London, 1967.*

(12) Kimura, K. Trenchless technology in Japan, *Proc. Osaka No-Dig Conference*, Osaka, 1990

(13) Rhodes, J. and Hagan R. Jacking sewer pipe in tunnel, *Proc. Concrete Pipe Engineering Seminar*, Lancing, Michigan, 1969.

(14) Craig, R. and Moss, A. *Pipe jacking - informal discussion*, British Tunnelling Society, May ,1983.

(15) Stein, D.et al. Small diameter guided pipes: development and experience from Hamburg, *Proc. Europipe '83 Conference*, Basle, 1983.

(16) Gyro compass guides curved pipe jacks, *Pipe Jacking News, Issue 7.*

(17) Shima, K. Tunnelling machine position analysis system, *Proc. No-Dig 90 Conference*, Osaka, Japan, 1990.

(18) Kuwahara H. et al Application of fuzzy theory to the control of shield tunnelling, *Proc. Third IFSA Congress*, Seattle, USA, August 1989.

(19) Takatsu T. et al On a common fuzzy control system for microtunnelling, *Proc. No-Dig Conference*, Paris, 1992.

(20) Tomohiro K. et al A new position detector system for microtunnelling using a mounted optical fibre gyroscope, *Proc. No-Dig Conference*, Paris 1992.

(21) Stein, D. et al *Microtunnelling*, Ernst & Sohn, Berlin, 1989.

(22) Richardson, M. and Scruby, J. The Unitunnel in action, *Tunnels and Tunnelling, Issue 13*, 1981.

(23) Anon A healthy spirit of innovation, *New Civil Engineer*, 15 October 1981.

(24) Meseck, H. Separation of supporting fluids for shield tunnelling, *Towards New Worlds in Tunnelling*, Balkema, 1992.

(25) Anheuser, L. Feststoffirennung aus Fordersuspensionen, *Proc. International Conference on Pipeline Construction*, Hamburg, 1987.

(26) Pezzimenti, A. and G. Microtunnelling in Australia, *Proc. No-Dig Conference*, Washington DC, 1988.

(27) Nishida, H. et al New thrust system 'SSK Method', *Proc. No-Dig 90 Conference*, Osaka, 1990.

(28) Najafi, M. et al Details of Kidoh/Iseki/Carlon research project conducted at TTC, *Proc. Trenchless Technology Seminar*, Vicksburg, 1993.

(29) Bristow, N. A variable diameter microtunnelling system, *Proc. No-Dig Conference*, Paris, 1992.

(30) Nojiri, Y. & Yamaguchi, S. A new microtunnelling method by ACE Mole, *Proc. No-Dig Conference*, Rotterdam, 1990.

(31) Husakabe, T. Muckless shield machine, *Proc. No-Dig 90 Conference*, Osaka, 1990.

(32) Nishino, T. et al Pipeline renewal by pipe jacking, *Pro. No-Dig 90 Conference*, Osaka, 1990.

(33) Geisler, P. Renewal of damaged sewers by hydraulic pipe jacking, *Proc. No-Dig 90 Conference*, Osaka, 1990.

(34) Uffman, H.P. Replacement laying of public and private house connection lines, *Proc. 2nd International Microtunnelling Symposium*, Munich, April 1992.

(35) Thomson, J. and Herbert, J. Trenchless design for the South Coast sewerage scheme in Barbados, *Proc. No Dig Conference*, Washington, 1992.

(36) Fujita, S. The legal system for road works in Japan, *Proc. No-Dig 90 Conference*, Osaka, October 1990.

(37) *New Roads and Streetworks Act*, HMSO, UK, 1991.

(38) Peters, D.C. Social costs of sewer rehabilitation, *Proc. Reading Conference*, UK, 1984.

(39) Castronova, J. *Proc. Pipeline Crossings Conference*, American Society of Civil Engineers, Denver 1991.

(40) National Academy of Sciences *Geotechnical site investigation for underground projects*, National Academy Press, 1986.

(41) Hair, C. Obstacle evaluation for crossing design, *Proc. Pipeline Crossing Conference*, American Society of Civil Engineers, Denver, 1991.

(42) Hair, C. Subsurface conditions affecting horizontal directional drilling, *Proc. Trenchless Technology Seminar*, Vicksburg, 1993.

(43) McLean, A.C. and Gribble, C.D. *Geology for Civil Engineers*, George Allen and Unwin, 1979.

(44) Kuroiwa, M. et al Development of a shallow seismic reflection survey for the microtunnelling method, *No Dig Conference*, Washington, 1992.

(45) Berosch, M. et al The use of ground penetrating radar on a microtunnelling site, *Proc. No Dig Conference*, Paris, 1992.

(46) Erling, J.C. and Lantier, F. Appraisal of electric and microgravimetric methods in underground works, *Proc. No Dig Conference*, Paris, 1992.

(47) Garnett, A. Electromagnetic location and monitoring technology, *Proc. No Dig 90 Conference*, Osaka, Japan, 1990.

(48) Koyabu, J. The development of Radarman, *Proc. No Dig Conference*, Paris, 1992.

(49) Henri, P. Pipes for microtunnelling: Suitability of use *Proc. No Dig Conference Proceedings*, Paris, 1992.

(50) Kondo, et al Advanced microtunnelling technology for telecommunication conduits, *Journal of Telecommunications Research*, October 1986.

(51) Berosch, M. and Revol, D. Trenchless technology for water supply, *Proc. No-Dig Conference*, Hamburg, 1991.

(52) Shah, D. & Jain S. Remotely controlled pipe jacking experiences in Oakwood Beach interceptor sewer Staten Island, *RETC Conference proceedings*, Seattle, 1991.

(53) Hiroharu, N. et al New thrust system "SSK Method", *Proc. No-Dig 90 Conference*, Osaka, 1990.

(54) Akiba, T. & Yamada T. The Enviliner microtunnelling system *Proc. No Dig 89 Conference*, London, 1989.

(55) Shibahara, S. The calculation of strength of PVC pipes *Proc. No-Dig 90 Conference*, Osaka, 1990.

(56) Marston, A. The Theory of external loads on closed conduits in the light of latest experiments, *Bulletin 96*, Engineering Experiment Station, Iowa State College, Ames, Iowa, USA, 1930.

(57) Spangler, M.G. The Structural design of flexible pipe culverts, *Bulletin 153*, Engineering Experiment Station, Iowa State College, Ames, Iowa, USA, 1941.

(58) O'Rourke, T.D., El-Gharbawy, S.L. and Stewart, H.E. Soil loads at pipeline crossings, *ASCE Specialty Conference on Pipeline Crossings*, Denver, March 1991.

(59) Chapman, D.N. *Ground movements associated with trenchless pipelaying operations (PhD thesis)*, Loughborough University of Technology, 1993.

(60) Rogers, C.D.F. and Yonan, S.J. Experimental study of a jacked pipeline in sand, *Tunnels and Tunnelling, Vol. 4, No. 6*, June 1992.

(61) *Statisiche Berechnun bon Vortriebsrohren*, Abwassertechnische Vereingigung, ATV, Germany, 1990.

(62) Ingraffea, A.R., O'Rourke, T.D., Stewart, H.E., Barry, A. and Crossley, C. Guidelines for uncased crossings of Highways and Railroads, *ASCE Specialty Conference on Pipeline Crossings*, Denver, March 1991.

(63) Stewart, H.E. and O'Rourke, T.D., Live loads for pipeline design at railroads and highways, *ASCE Specialty Conference on Pipeline Crossings,*, Denver, March 1991.

(64) Watkins, R.K. and Spangler, M.G. Some Characteristics of the modulus of passive resistance of soil. A study in similitude, *Proc. HRB, Vol. 37*, Washington DC, USA, 1958.

(65) White, H. Precast Concrete Jacking Pipes: Man-entry to micro, *Proc. Pipejacking and Microtunnelling Conference*, London, 1991.

(66) Möhring, K. Contributions of the contract placing authority to the economic efficiency of microtunnelling technology, *Proc. 2nd International symposium on Microtunnelling*, Munich, 1992.

(67) *Code of practice for microtunnelling*, Yorkshire Water Plc., Second Edition, 1991.

(68) *AREA specifications for pipelines*, American Railway Engineering Association USA, 1965.

(69) Pipeline crossings of railroads and highways, *Interim specifications*, American Society for Civil Engineers (ASCE), 1964, USA.

(70) Crossings of petroleum pipelines, *Standard Z 183*, Canadian Standards Association, Canada.

(71) Crossings of gas pipelines, *Standard Z 184*, Canadian Standards Association, Canada.

(72) Kainuma, N. et al Development of a new pipe jacking method for small bore power conduits, *Proc. No-Dig Conference*, Hamburg, 1991.

(73) Terzaghi, Geologic Aspects of Soft Ground Tunnelling, *Chapter 11, Applied Sedimentation* edited by P Transk, John Wiley and Sons, New York, 1950.

(74) Smirnoff, T. P. Tunnelling In Soft Ground, *Chapter 12, Underground Structures, Design and*

Instrumentation, edited by R S Sinha, Elsevier, Amsterdam, 1989.
(75) Chapman, D. N. and Rogers, C. D. F. Ground Movements Associated With Pipe Installation and Tunnelling, *Proc. 4th International Conference on Ground Movements and Structures p. 6.1-6.17*, Cardiff, July, 1991.
(76) Atkinson, J. H. and Mair, R. J. Soil Mechanics Aspects of Soft Ground Tunnelling, *Ground Engineering, Vol. 14, No.5, p.20-28*, 1981.
(77) Potts, D. M. *Behaviour of Lined and Unlined Tunnels in Sand (PhD Thesis)*, Cambridge University, UK, 1976.
(78) Norris, P. *The Behaviour of Jacked Concrete Pipes During Pipe Installation (D Phil Thesis)*, Oxford University, UK, 1992.
(79) Peck, R. B. Deep Excavation and Tunnelling in Soft Ground", State-of-the-Art Volume, Proc. of 7th Int. Conf. on Soil Mechanics and Foundation Engineering, Mexico City, p.225-290, 1969.
(80) Broms, B. B. and Bennermark, H. Stability of Clay at Vertical Openings, *Journal of Soil Mechanics and Foundation Div. ASCE, Vol. 93, p. 71-94*, SM1, 1967.
(81) Klein, S. J. Geotechnical Aspects of Pipe Jacking Projects, *Proc. ASCE Specialty Conference on Pipeline Crossings, p 113-128*, Denver, March 1991.
(82) Mair, R. J. *Centrifugal Modelling of Tunnel Construction in Soft Clay (PhD Thesis)*, Cambridge University, UK, 1979.
(83) Mair, R. J. Discussion on Tunnel Face Pressures in Soft Clay, *Proc of 8th Asian Regional Conference on Soil Mechanics and Foundation Engineering, Vol. 2, P.290, 1987.*
(84) Chapman, D. N. *Ground Movements Associated with Trenchless Pipelaying Operations (PhD Thesis)*, Loughborough University of Technology, UK, 1993.
(85) Attewell, P. B. and Yeates, J. Tunnelling in Soil, *Chapter 6, Ground Movements and Their Effect on Structures*, edited by Attewell P B and Taylor R K, Surrey University Press, 1984.
(86) O'Reilly Michael P and Rogers C D F (1990), Ground movements associated with pipejacking and trenching, *Proc. of No Dig 90*, Rotterdam, April, p. B1-B8.
(87) Attewell, P. B., Selby, A. R. and Yeates, J. *Soil Movements Induced by Tunnelling*, Blackie, Glasgow and London. Published in the USA by Chapman and Hall, New York, 1986.
(88) O'Reilly, Myles P. and New, B. M. Settlements Above Tunnels in the United Kingdom - Their Magnitude and Prediction, *Proc. of Tunnelling '82 Symposium*, IMM, London, 1982.
(89) Sagaseta, C. Analysis of Undrained Soil Deformation due to Ground Loss, *Geotechnique, Vol. 36, No.3*, September. 1987
(90) Glossop, N. H. *Soil Deformations caused by Soft Ground Tunnelling (PhD Thesis)*, University of Durham, UK, 1977.
(91) Lasudry, R. & Delbaere, D. Cooling water for Tihange, *World Tunnelling*, April 1990.
(92) Tomlinson, M.J. *Foundation design and construction*, Pitman International, 1983.
(93) Koerner, R.M. *Construction and geotechnical methods in foundation engineering*, McGraw Hill, 1985.
(94) Nussbaumer, M. & Berger, K. Pipe jacking water saturated ground, *World Tunnelling*, April 1989.
(95) Anon, Hydraulic thrust jacking for the construction of a utility line in Nice, *Westfalia Becorit literature.*
(96) Herzog, M. Die Pressenkräfte bei Schildvortrieb und Rohrvorpressung im Lokkergestein", *BMT, No. 6, pp. 236-238*, 1985.
(97) Scherle, M. Rohrvortrieb, *Part 2, Bauverlag*, Wiesbaden/Berlin, 1977.
(98) Standard Pipe Jacking Construction System, *Manual of Iseki Poly-Tech Inc.*
(99) Craig, R. N. Pipe Jacking: A State-of-the-Art Review, *Technical Note 112*, CIRIA, London, 1983.
(100) O'Reilly, M.P. & Rogers, C. Pipe jacking forces, *Proceedings of International Conference on Foundations and Tunnels*, Edinburgh, 1987.
(101) Anon, Long Distance Driving Method for Ductile Iron Pipe, *Water Services, pp 516-522*, October 1981.
(102) Haslem, R. F. Pipe Jacking Forces: from Theory to Practice, *Proc. of Infrastructure, Renovation and Waste Control, North West Association Centenary Conference, pp 173 - 180*, Institution of Civil Engineers, 1986.
(103) Roark, R. J. and Young, W. C. *Formulas for Stress and Strain*, McGraw Hill, London and New York, 1976.
(104) Peck, R. B. Deep Excavation and Tunnelling in Soft Ground, *Proc. of 7th International Conference on Soil Mechanics and Foundation Engineering, pp 225-290*, Mexico City, Mexico, 1069.
(105) Mair, R. J. *Centrifugal Modelling of Tunnel Construction in Soft Clay, (PhD Thesis)*, University of Cambridge, 1979.
(106) Atkinson, J. H. and Bransby, P. L. *The Mechanics of Soils: An Introduction to Critical State Soil Mechanics*, McGraw Hill, London, 1978.
(107) Chapman, D. N. *Ground Movements associated with Trenchless Pipelaying Operations (PhD thesis)*, Loughborough University of Technology, 1993.
(108) Yonan, S. J. *Pipe Jacking Forces in Sand (PhD thesis)*, Loughborough University of Technology, 1993.
(109) Norris, P. and Milligan, G.W.E. Pipe End Load Transfer Mechanisms During Pipe Jacking, *Proc. of Int. Conf. on Trenchless Technology, No Dig 92. Paper H3*, Washington DC, USA, 1992.
(110) Rogers, C. D. F., O'Reilly, M. P. and Atkin, R. Pipe Jacking Beneath Burnham-on-Sea: A Case History, *Proc. of Int. Conf. on Foundations and Tunnels 89*, Engineering Technics Press, Edinburgh, 1989.
(111) Rogers, C. D. F., O'Reilly, M. P. and Atkin, R. Prediction of Pipe Jacking Forces Based on Site Observa-

tion, *Proc. of Int. Conf. on Foundations and Tunnels 89*, Engineering Technics Press, Edinburgh, 1989.

(112) Takeshita, S. The Effectiveness of Friction Cut Method and Bentonite Suspension Injection Method in Pipe Jacking, *Proc. of Int. Conf. on Foundations and Tunnels 89*, Engineering Technics Press, Edinburgh, 1989.

(113) Ishibashi, N. Japan's Recent Small Diameter Pipe Jacking Construction Methods, *Proc. of Third International Conference on Trenchless Technology, No Dig 88*, Washington DC, 1988.

(114) Norris, P. and Milligan, G.W.E. Frictional Resistance of Jacked Concrete Pipes at Full Scale, *Proc. of International Conference on No Trenches in Town, No Dig 92, pp 121-128*, Paris, 1992.

(115) Guilloux, A. and Legaz, C. Synthesis of Microtunnelling Projects in the Val de Marne, *Proc. of International Conference on No Trenches in Town, No Dig 92, pp 151-158*, Paris, 1992.

(116) Toyhama, S. Pipe jacking with the use of membrane wrapping, *Proc. of No-Dig 87 Conference*, London, 1987.

(117) Toyhama, S. Case study on pipe jacking with membrane wrapping, *Proc. of No-Dig 88 Conference*, Washington DC, 1988.

(118) Rogers, C. D. F. and Yonan, S. J. Experimental Study of a Jacked Pipeline in Sand, *Tunnels and Tunnelling, Vol. 24, No. 6, pp 35-38*, June 1992.

(119) Stevens, W. W. Ductile Iron Jacking Pipes, *Proc. of Fourth Int. Conf. on Trenchless Construction for Utilities, No Dig 89, pp 2.2.1-2.2.10*, London 1989.

(120) Ripley, K. J. *The Performance of Jacked Pipes, (DPhil Thesis)*, Oxford University, 1989.

(121) Nagata, N., Kanari, H., Yamamoto, T., Okano, M. and Tsushima, K. Design of Stress-Absorbing Plates for Pipe-Jacking Method, *Proc. of Sixth International Conference on Trenchless Construction for Utilities, No Dig 90*, Osaka, October 1990.

(122) Milligan, G.W.E. and Ripley, K. J. Packing Material in Jacked Pipe Joints, *Proc. of Fourth International Conference on Trenchless Construction for Utilities, No Dig 89, pp. 2.1.1 - 2.1.10*, London, 1989.

(123) Boot, J. C. and Husein, N.M. Vitrified Clay Pipes Subject to Jacking Forces, *Proc. of Conference on Pipejacking and Microtunnelling 91*, Pipe Jacking Association, London, 1991.

(124) Nussbaumer, M. and Berger, K. Pipe Jacking in Water- Saturated Ground, *World Tunnelling Special Edition*, April 1989.

(125) Atkin, R. Mechanical Curved Pipe Jacking by Gyro Compass, *Proc. of Conference on Pipejacking and Microtunnelling 91*, Pipe Jacking Association, London, 1991.

(126) *Occupational Safety and Health Administration (OSHA) Act Subpart*, 1926.

(127) Cruz, E. Microtunnelling under difficult conditions in New York City, *Proc. Trenchless Technology Seminar*, Vicksburg, 1993.

(128) Taylor, A. & Brown, M. Entry and exit portals for tunnel boring machines in unstable soils, *Proc. Pipejacking and Microtunnelling Conference*, London, 1991.

(129) *SIA Norm 195*, Schweizerischer Ingenieur und Architekten (SIA) Verein Pressevortrieb.

(130) Clarkson, T. and Thomson, J. Pipe jacking state of the art in UK and Europe, *ASCE Journal of Transportation Engineering*, January, 1983.

(131) Thomson, J. Horizontal earth boring, *Proceedings of the Institution of Civil Engineers*, April, 1967.

(132) Harding, H. The choice of expedients in civil engineering, *ICE Works Division Proceedings*, 1946.

(133) *ICE Conditions of Contract, 6th Edition*, The Institution of Civil Engineers, London.

(134) *Impact of various contract types and clauses on project performance*, Construction Industry Institute, University of Texas, 1986

(135) *Avoiding and resolving disputes during construction*, American Society of Civil Engineers, 1991.

(136) *Conditions of Contract*, Institution of Chemical Engineers, London.

(137) Neale, R.S. Managing safety at roadworks - EC directives, *Proceedings HAUC 91*, Birmigham, 1981.

(138) Medical code of practice for work in compressed air, *CIRIA Report 44*, Construction Industry Research and Information Association, 1982.

(139) *UVV Safety regulations for construction work underground*, Tiefbau-Berufsgenossenschaft, Munich, 1986.BS6164 "Safety in Tunnelling" British Standards Institute 1990,

(140) *VBG 37, The safety rules for pipeline construction work*, Tiefbau- Berufsgenossenschaft, Munich, 1986.

(141) Teague, F. and Lau, K. Control and management of utility works in Hong Kong, *No-Dig 90 Conference Proceedings*, Osaka, 1990.

(142) Balasubramaniam, K. No-dig sewers in Singapore by microtunnelling methods, *No-Dig 88 Conference Proceedings*, Washington DC, 1988.

(143) *Jacking pipe specification*, Japanese Sewage Works Agency, JSWAS A-2, 1973.

(144) *Regulations for Pipe Jacking Operations*, ATV DIN 18 319, Abwasser Technische Vereinigung, 1992.

(145) *Static calculations of driven pipes*, ATV-Standard A 161/GW 312, Abwasser Technische Vereinigung, 1990.

(146) Friede, Dr. H., Quality Control with trenchless construction, *No-Dig Conference Proceedings*, Hamburg, 1991.

(147) Piquereau, G. Automates de pilotage de tunneliers, *Colloque Internationale, Tunnels et Microtunnels en Terrain Meuble*, Paris, 1989.

(148) Kamber P. Microtunnelling through screes and river gravels in Switzerland, *Proceedings of the 2nd International Microtunneling Symposium*, Munich, 1992.

(149) Rippl and Krautsschun Isoschaum foam for pipe jacking in solid rock, *Tunnel Volume 3*, 1987.

(150) Confined space standards, *Standard 29CFR 1910.146*, Occupational Safety and Health Administration (OSHA), 1993.

(151) Brierley et al Toxic tunnelling, *World Tunnelling*, November 1991.

(152) Yorkshire prefers the trenchless option, *Underground*, 1989.

(153) Coe, G. and Webb, S. Signing standards at roadworks, *Proc. PTRC Conference*, 1992.

(154) *Trenchless construction of pipelines, Report prepared for Transport Research Laboratory 1993*, Jason Consultants.

(155) Moss A. Microtunnelling for contractors, *Proc. No Dig Conference proceedings,* Rotterdam, 1989.

(156) Elvidge, A.F. The potential for microtunnelling in the UK, *No-Dig Conference proceedings*, London, 1987.

(157) Clarke, I. Micro utilisation on green field development, *No-Dig International*, December, 1992.

(158) Dip-Ing Möhring Mehr als 100,000 m microtunnelbau in Berlin, *Wasser Berlin Conference proceedings*, 1993.

(159) Young, O.C. and O'Reilly, M.P. *A guide to design loadings for buried rigid pipes*, HMSO, London, 1983.

(160) Rogers, C.D.F. and Chapman, D.N. Ground Movements caused by Pipejacking Operations, *Ground Engineering*, Vol. 27, No.8, 1994.

(161) DeMoor, E.K. and Taylor, R.N. Ground Response to a Sewer Tunnel in Very Soft Ground. *Proc. of 6th Int. Symp. on Tunnelling 1991*, London, published by Elsevier. pp.43-54

Glossary

Annular space	(a) The overcut void formed by the outside diameter of the shield or head around the pipe or casing. (b) The ring-shaped space between the internal diameter of the existing pipe and the external diameter of a new sliplined pipe.
Articulated head	A shield or cutting head which is in two sections allowing the front section to be angularly deflected.
Auger boring	The technique of forming a horizontal hole by cutting the soil with a rotating auger head and removal of muck by helically wound screw conveyors. A casing is normally simultaneously installed by jacking.
Bentonite	A natural clay material having thixotropic properties which is used as a basic ingredient for drilling muds and lubricants.
Berlin system	Property laterals radially connected to main or subsidiary access shafts.
Butt-ended	Square-ended pipe ends presenting maximum contact area.
Caisson shaft	Shaft constructed by preforming the structure at the surface and then sinking it into the ground by excavating.
Casing	Usually a steel tube bored or driven to provide a duct into which the permanent pipe is placed.
Clayware	General term for vitrified clay pipes.
Closed shields	Shields which incorporate a closed pressure chamber and do not allow direct access to the face.
Cone Penetration Test	CPT is an in-situ method of testing soil using a Static or Dutch Cone.

Conventional shields	Shields which allow some form of direct access to the soil face. Also known as open shields.
Crossing	A section of utility line where a pipe or cable has to pass under an existing road, railway, river or other area where the surface should not be disturbed.
Cutting head	The device at the leading end of the shield which cuts the soil face.
Deformation	Soil movement arising from the work which can cause settlement or heave at the surface.
Deviation	The variation of line and level from that specified.
Dewater	Reduction of the ground water level by pumping. Often used in connection with the vacuum wellpoint method.
Drive pit	Also known as a jacking pit or drive shaft, it is the temporary shaft sunk on the line from which the work is undertaken.
Earth pressure balance	The application of pressure to counter the movement of soil into the shield.
Eye	Also known as a portal or window, it is a temporary seal device placed in the shaft wall of drive and reception pits to allow the shield and the pipes to enter and exit the ground without soil, water and slurry flowing into the working pits.
Face	The undisturbed soil immediately in front of the head.
Ground Probing Radar	Also known as Ground Penetrating Radar and Ground Scanning, GPR uses electromagnetic radiation pulses and the measurement of the reflections to investigate what lies below the ground surface.
Grade	A term particularly used in the USA to denote the levels to which a pipe is laid.
Greenfield site	Construction work in a previously undeveloped area.
Gripper System	A method of transmitting the jacking load onto the wall of plastic pipes as they are installed.
Inclinometer	An instrument which measures direct or remote angular tilt of a shield or cutting head.

Intermediate Jacking Station	A series of hydraulic jacks enclosed within an outer steel casing placed in the line between the butt ends of two special pipes. By pushing the forward section of pipes against the reaction of those behind, additional jacking load can be added without increasing the maximum stress on the pipes.
Jacking pipe	The pipe used for lining a bore where it is added in the drive pit and pushed forward.
Joint	The sealing and joining arrangement betwen two pipes. For jacking work joints are of the rebated or sleeved type.
Laterals	Sewer house connections.
Location	Identification of the position of underground pipes, cables and other objects using electromagnetic, acoustic, radar and other devices.
Magnetometer	A device for measuring the magnetic field.
Membranes	Also known as drag sheets, membranes are sheets of material or jackets placed between the soil and the pipe being jacked to improve the interface and reduce friction.
Microtunneller	A small (usually < DN 900 mm) tunnelling system that is remotely monitored and steered.
Mud	The formulation of water and additives which are prepared and pumped into the pressure chamber. [Cf. Slurry]
N value	Values from the Standard Penetration Test based on the number of blows required to drive a standard sampler 12" to provide a measure of soil strength.
Overcut	The difference between the diameter of the shield and the pipes being jacked.
Pilot boring	A two-pass microtunnelling method in which a pilot bore is driven initially and a second reaming pass then enlarges it to the required bore size.
Piezocone	An insitu soil testing instrument mainly used for testing pore water pressure.
Pipejacking	Also known as pipe pushing. Term for (a) the principle of installing pipe sections to form the lining behind a cutting head or shield, and

	(b) the technique of installing pipe sections (DN 900 mm and greater) by pushing in sections.
Pipe replacement	A method of simultaneously excavating and removing the old pipe and replacing with a new one using a remote-control microtunneller. Sometimes called pipe-eating.
Pressure balance	A classification of tunnel boring machines which provide a positive pressure (soil, water or slurry) to resist active and passive soil pressures and water head.
Reaction wall	Also known as a thrust wall. A structure designed to distribute the loads from jacking into the adjacent soil.
Reception shaft	The excavation or pit which is at the end of the drive.
Rehabilitation	The whole range of alterations and additions to an existing network to improve the service.
Renovation	The structural repair and improvement of an existing service.
Replacement	The installation of a new pipe on the line of the old service.
Rock Quality Designation	A percentage ratio based on core recovered against length of core drilled.
Seismic ground investigation	An investigation technique which measures with a seismograph the speed of propogation of artificially induced vibrations through differing rock and soils. Refraction methods are most commonly used in relatively shallow investigation.
Separation plant	Arrangements for separating the spoil from the slurry.
Single pass	Also called one pass. The method of installation where the permanent pipe is installed simultaneously with the boring. [Cf. two pass]
Skin friction	The contact forces of the soil when pushing a pipe into a bore.
Slime shield	A particular form of earth pressure balance shield which mixes the soil at the face with additives to produce a consistent plastic material.
Slip coatings	Temporary or permanent coatings applied to the outside of a pipe to reduce skin friction.

Slurry	A general term used for bentonite/water muds and also for clay/water mixes which can be pumped. The correct usage is for mud which has been polluted with cuttings.
Slurry TBM	A mechanical shield which incorporates a counterbalance slurry chamber.
Surface heave	Disturbance and deformation at the ground surface created by positive soil displacement.
TBM	A general term for any tunnel boring machine.
Thrust boring	Also known as Rod-pushing. A basic method of forming a hole by soil displacement. A closed-end pipe or head with connecting rods is jacked into the soil.
Trenching	The traditional method of laying services in an open-cut excavation.
Trenchless	The generic description for all installation, replacement and renovation systems that do not require open-cut trenching.
Two pass	A two-stage method of installation that involves driving a pilot bore or temporary casing and then installing a permanent pipe while pushing out the pilot or temporary casing.
VCP	Vitrified clay pipe.
Water table	The natural level of water in the soil.

Index

FSC
www.fsc.org
MIX
Papier | Fördert
gute Waldnutzung
FSC® C083411

Kolibri 360 GmbH
Ferdinand-Jühlke-Straße 7
99095 Erfurt, Deutschland
produktsicherheit@kolibri360.de